Computational Music Science

Series editors

Guerino Mazzola
Moreno Andreatta

More information about this series at https://link.springer.com/bookseries/8349

Nicolas Namoradze

Ligeti's Macroharmonies

A Graphical-Statistical Analysis of Book 3
of the Piano Etudes

 Springer

Nicolas Namoradze
New York, NY, USA

ISSN 1868-0305   ISSN 1868-0313   (electronic)
Computational Music Science
ISBN 978-3-030-85696-0   ISBN 978-3-030-85694-6   (eBook)
https://doi.org/10.1007/978-3-030-85694-6

Mathematics Subject Classification: 00A65, 42A38, 62P99

This Springer imprint is published by the registered company Springer Nature Switzerland AG.
The registered company address is: Gewerbestrasse 11, 6330 Cham, Switzerland

Preface

I first met György Kurtág in the spring of 2006. Thirteen years old at the time, I had recently begun my piano studies at the precollege program of the Liszt University of Music in Budapest; I was preparing a short program of Kurtág's music for a performance as part of a series of concerts celebrating his 80th birthday, and my piano professor arranged for me to meet and play these pieces for the composer himself while he was in town.

We visited the Kurtágs at their apartment on a rainy March afternoon. After working on the program with the composer (an experience I remember vividly), he and his wife—the late, great pianist Márta Kurtág—invited us to stay a little longer for tea and cakes. At some point the conversation briefly turned to another composer. "I recently visited my friend György Ligeti," György Kurtág remarked, "he is dying."

While I had heard of this other great living Hungarian composer, I was not yet familiar with his music. Yet something about Kurtág's remark and the reverence with which he mentioned Ligeti's name became immediately stuck in my consciousness, and I soon began exploring his oeuvre. A few months later Ligeti passed away in Vienna. By the next year, I was immersing myself in Ligeti's piano etudes and would soon be performing a selection of them extensively (they had in fact become my calling cards as a pianist). Among the etudes that were at the forefront of my programs were the four final ones that constituted Book 3, pieces that I had developed a particular attachment to.

Thus, my fascination with Book 3 has spanned more than a decade—the better part of my life as a musician. These four etudes are arguably Ligeti's most mysterious: they are rarely performed or recorded, and there is virtually no literature dedicated to them (despite the many writings on his earlier etudes). In this book, I set out to explore what is one of their most remarkable compositional features and do so with the hope that this analysis may encourage greater interest in these works that not only are masterfully wrought and, in many ways, radical, but also serve as revealing capstones to a compositional oeuvre that is unique in the extent and nature of its stylistic developments.

New York, NY, USA   Nicolas Namoradze

Acknowledgments

This monograph originally appeared as my doctoral dissertation at the Graduate Center, City University of New York. I owe a great debt to my advisor Joseph Straus, whose knowledge and creative approach have been an inspiration in this process. Norman Carey's guidance was invaluable during the writing of this work, as well as throughout my years at the Graduate Center. In addition, his support during the publication of this work has been instrumental. I would like to thank the three further members of my defense committee, who made important contributions: David Schober for his helpful suggestions and eagle-eyed reading of the manuscript, Jeff Nichols for his perceptive discussions on the topic, and Lukas Ligeti for his unique insights on his father's life and music.

I am very grateful to the editors of Springer's Computational Music Science series, Guerino Mazzola and Moreno Andreatta, for their support of this project and guidance throughout this process, and to the Springer production team for the preparation of the manuscript for publication. Many thanks to Dagmar Schütz-Meisel and the staff at Schott Music for their kind permission to reproduce score excerpts from Ligeti's etudes in this book. Finally, I am indebted to my parents, without whose support during my studies this project would not have been possible.

Contents

Chapter 1
Introduction

1.1   New Directions in Book 3

Let us begin by observing the opening page of the first etude of Book 3 of Ligeti's piano etudes, Etude 15, *White on White* (Fig. 1.1). There is no varying of rhythm: it is a string of uninterrupted half-notes. The *legato* articulation is constant—as is the volume, given that there are no accents or dynamic changes. There is a general homogeneity of texture as most simultaneities consist of three notes, with the occasional dyad and tetrachord. The register is relatively constant, the passage occupying the central area of the keyboard. Importantly, the two hands are in canon: the left hand is an exact repetition of the right an octave below and a single pulse later.[1] There is repetition in thematic material as well, given that bars 6–10 reiterate and reharmonize the melody presented in the first five bars. (In fact, the following bars 11–15 are an almost literal repetition of bars 1–5, with a written-out *ritardando* at the end.) Finally and crucially, the passage is entirely diatonic: there is no departure from the strict seven-pitch class framework, the white-note set.[2]

The combination of such extreme minimalism in these parameters is a significant departure from Ligeti's earlier work in the genre. This opening of *White on White* in fact foreshadows many of the processes that will take place in the rest of the etudes of Book 3. Ligeti uses repetition of material frequently, even writing a repeat sign in the final etude—the first and only appearance of this marking in his etudes. With only a few exceptions, all passages in these last four etudes completely adhere to an even rhythmic framework, which in faster tempi creates a *perpetuum mobile* texture.

---

[1] The right and left hands will henceforth be referred to as RH and LH.

[2] Pitch class is hereafter abbreviated to PC.

N. Namoradze, *Ligeti's Macroharmonies*, Computational Music Science, https://doi.org/10.1007/978-3-030-85694-6_1

Fig. 1.1 Etude 15, bars 1–10, score. © 2005 Schott Music, Mainz—Germany

While this in itself is not new to Ligeti, in previous etudes the rhythmic values of this even, constant motion would serve as a lowest common denominator for polyphonic, polyrhythmic layering, something that is largely absent in Book 3; while there are certainly passages with accents, in the vast majority of them they are either coordinated in the two hands or displaced by a canon-relationship between them. The extensive and strict use of canon is in itself also a new feature in these final etudes. This greater dynamic homogeneity and interconnectedness of the two hands contributes to the relative textural simplicity and purity of these works—something that is also reflected by the complete dispensing of barlines in the etudes following *White on White*. While Ligeti had often included a disclaimer in his earlier etudes that barlines are merely for orientation (as he does for the dotted barlines in *White on White*) he never did away with them entirely; the more limpid and rhythmically straightforward textures of the final etudes on the other hand no longer necessitate this "grid."[3]

However, what is arguably the most significant restriction is that of pitch class collection—generally referred to in this book by the simpler term "scale." *White on White* will continue this strict adherence to the white-note set through the end of the etude, at the conclusion of which there is a light smattering of other, black notes; there is never anything resembling a true movement away from this diatonic focus on one scale area. This is a radical departure from his earlier work. Save for some portions of one etude (the tenth, *Der Zauberlehrling*—"The Sorcerer's Apprentice"), Ligeti's previous etudes are chromatically saturated throughout; those that use more restricted scales do so in certain textural layers, which, when superimposed upon other simultaneously occurring layers, yield the full chromatic set. On the other hand, in the rest of the etudes of Book 3 not only do *complete* textures occupy certain clearly-defined scales, but Ligeti often juxtaposes several such restricted scale areas along the course of the etudes—modulating between them as common-practice music would modulate between keys—and uses chromatic saturation only as a destabilizing counterpart to the steadiness of the cleaner passages.

The term this book uses for the general scale area that a passage might adhere to is "macroharmony." The term is adopted from Dmitri Tymoczko's "Geometry of Music" (2011), which originated it and set out some key parameters for understanding this concept. However, the use of the term here differs somewhat from Tymoczko's: he defines macroharmony as "the total collection of notes used over small stretches of musical time" (Tymoczko 2011, 15), while here a certain passage may embody a particular macroharmony even if not all of the notes in the passage belong to it. To illustrate, the macroharmony in a later passage in *White on White* shown in Fig. 1.2 would here still be considered the white-note set, even though there are two black-note intrusions in bar 40. This means that, rather than the PC set that defines the macroharmony itself changing, it is the correlation (i.e. the strictness of the adherence) to it that has shifted; the macroharmony has become fuzzier.

This monograph seeks to show that in Book 3 of the piano etudes, Ligeti uses macroharmonic processes to articulate structures and narratives, and that macroharmony is often the central feature of a work that defines its dramatic arc.[4] The primary scale type used in such procedures is the diatonic. Importantly, the processes are largely reflective and governed by structures inherent in the diatonic scale, the most significant of which are fifth-relationships; this leads to relations and

---

[3]For a discussion of the relative usefulness (or lack thereof) of this system of barlines, see Talgam (2019).

[4]The use of terms such as drama and narrative is entirely abstract, as there is no concrete plot or story in the etudes; it rather refers to the expressive arc that each work presents.

Fig. 1.2 Etude 15, bars 38–41, score. © 2005 Schott Music, Mainz—Germany

juxtapositions of macroharmonies that are often functional in nature, an important departure from the manner in which diatonicity is treated in his earlier etudes.

More generally, the analytical techniques developed in this book seek to shed light on our perception of varying levels of stability and instability brought about by the treatment of macroharmony and the calibration of its variables. Not only do scale types inherently differ in their degree of stability, but a listener's sense of "steadiness" may vary even in passages adhering to the same macroharmony due to factors such as pitch and PC centricity, interval content of the verticalities and correlation, all of which play an important function in the communication of dramatic narratives in the etudes of Book 3. The analytical methods used here attempt to in a sense quantify these levels of stability, developing a number of metrics that can be used to chart their change—and by extension approach a reflection of a listener's experience of this feature—over the course of a work.

Despite the extent of the stylistic departures in Book 3 and their significance as Ligeti's final works (the last etude, *Canon*, is his final composition) there is until now no (publicly accessible) literature devoted to their analysis.[5] Indeed, many of these features, especially the treatment of scale, is particular to these etudes: while Ligeti's other compositions of the final years (the *Hamburg Concerto* and the cycle for voice and percussion *Síppal, Dobbal, Nádihegedűvel*) also to some extent display this stylistic shift, it is nowhere near as pronounced as in Book 3. Why have these pieces, which represent the conclusion of the work of a composer whose stylistic transformations have been the subject of such extensive discussion, received such scant

---

[5] While Ligeti made revisions to the *Hamburg Concerto* (1998–1999) after the composition of *Canon* in 2001, he produced no new works until his death in 2006.

attention? Thus, before discussing the methods used in the present analysis at greater length, it is first important to address the compositional context of Book 3 in the "late style" of Ligeti's oeuvre and the discourse that surrounds it.

1.2   Ligeti and "Late Style"

Upon completing his Horn Trio in 1982, Ligeti remarked:

> Let me say that this Horn Trio is the first piece in this new Ligeti style... I am myself, but let us call this my last period, the period of my old age, I do not know how long it will last. (Szigeti 1983)

This comment is part of a pattern of statements from the composer from the 1980s onwards that emphasize stylistic breaks in his compositional development, partitioning his oeuvre into three periods: the first lasting until his escape from Hungary in 1956 and the second until the completion in 1977 of his opera *Le grand macabre* (Shaffer 2011, 205), which was followed by a few relatively fruitless years beset by illness before his finding of a "new style" (Szigeti 1983).

Ligeti himself described his new approach as something of a "third way," a rejection of both an avant-garde that has become "academic" and neo-expressions that "chew over an outmoded style" (Ligeti et al. 1983, 123); "I am trying to develop a harmony and melody which are no genuine return to tonality, which are neither tonal nor atonal but rather something else, above all in connection with a very high degree of rhythmic and metric complexity" (Bossin 1984, 238). This neither-nor idea remains a theme throughout his statements on his late style; more than a decade later, in the liner notes for a disc of his piano etudes (which included the then-newest one, *White on White*), Ligeti describes his music as "diatonic... and not yet tonal" (Ligeti 1996, 11).

The ambivalent, unique nature of this late style has been the subject of a considerable amount of scholarship. Beyond more general discussions of Ligeti's relationship to broader musical trends and the subject of "postmodernism" in his late style (see Floros 1996, Searby 1997, Steinitz 2003 and Toop 1999), analyses of his compositions from this period—and of the piano etudes in particular, the most significant set of works written during this time—focus variously on this new "rhythmic and metric complexity" (see Cuciurean 2000 and Taylor 2012), the influences of non-Western musical idioms (Bauer 2011, Scherzinger 2006 and Taylor 2003) and Ligeti's treatment of harmony (Drott 2003, Schaffer 2011, Searby 2001, Taylor 1994 and Quinnett 2014).

While it is this last category that is most relevant to the current investigation, it is nevertheless only tangentially related. In the current literature there is little dedicated discussion of Ligeti's use of scale and scale areas, and when the topic is occasionally touched upon it is usually used as an illustration of Ligeti's success in restoring "traditional" parameters such as melody, triads and diatonic scales while still occupying a middle ground between tonality and atonality. In the case of scale this is achieved through the superimposition of different scale types in the various layers—"combinatorial tonality" (coined in Steinitz 1996, 8; see also Callender 2017, 2)—the sum of which result in the full chromatic set, thus meaning that Ligeti can have it both ways (see, for example, Drott 2003, 294 and Hentschel 2006, 91). These conclusions are largely an extension of Ligeti's own statements on the matter.

However, the ramifications of the use of diatonic scales not as registrally distinct subsets of the chromatic gamut but rather as the *sole* scale area of a texture—as is the case in Book 3—have not yet been addressed. Given the by-now accepted wisdom of the traits of Ligeti's late style, the etudes of Book 3 have been lumped in under the umbrella of this neither-nor interpretation, without a dedicated examination of the manner in which the unique processes in these works may undermine it. The central arguments posited in this book include a claim that the procedures in Book 3 depart from those in Ligeti's previous etudes (and, more generally, his other mature works), particularly with regards to the increasingly functional treatment of scale area; the final etudes approach harmonic and formal processes that are increasingly "tonal." This is an idea that goes against not only much of the literature on Ligeti's late style, but also the composer's own statements.

There are two important caveats here. First, Ligeti's far more infrequent interviews from the period beginning in the late 1990s (presumably due to deteriorating health) leaves us with a dearth of first-hand insight regarding his compositional process in the final years. Another is the question of the accuracy of Ligeti's statements when discussing his music, and whether his characterizations were influenced by other considerations given the manner in which the composer wished to establish his legacy and influence the discourse around his work. This issue is discussed at length by Shaffer, who posits that Ligeti's descriptions of his late works as removed from the past were largely driven by how he wished to portray his "late style," even though these are perhaps not entirely precise reflections of the music itself:

> The conflict between Ligeti's developing concept of musical form that began in the late 1950s and his legacy-building project that began in the early 1980s is clear. The former would emphasize longer works with coherent syntactic structures, as well as an active awareness of the historical significance of musical elements. The latter would emphasize shorter or fragmented works that break from the now-exhausted material of the past... It is clear that in the latter part of his career, in spite of the fact that he continues to write music in line with his earlier rhetoric on form and syntax, Ligeti desires to be seen as a "late" composer—both in terms of his own career, and in terms of the broader history of music. Thus, while composing music that draws heavily on both tonal and atonal musics of the past, he states that his music is "neither tonal nor atonal." (Shaffer 2011, 209)

Shaffer's book on Ligeti's triadic harmony is in fact the first analysis to challenge the "neither tonal nor atonal" notion, concluding that "Ligeti composed meaningful harmonic successions... [that] have a strong relationship with some fundamental aspects of the successions and syntax of common-practice tonal music" (Shaffer 2011, 193). The present analysis hopes to further this reevaluation of the traits of Ligeti's late style through the discussion of macroharmony in Book 3.

Another issue regarding the position of Book 3 in the discussion of Ligeti's late style is its dismissal as simplistic—and perhaps less "interesting"—compared to the more complex earlier etudes. For example, while Steinitz admits that "the concentration in Book 3 on white notes and a gently inflected modality is unprecedented in Ligeti's music," he also characterizes them as "gentler and less dramatic than their predecessors, in fact more of a consolidation... This leaves open the question of whether Book 3 can achieve the stature of the other two" (Steinitz 2003, 313); the novel macroharmonic processes are seen as a placid crystallization of previous ideas in Ligeti's music, rather than a significant change in approach. Through the discussion of the unique features of these final etudes, this book seeks to shed some light this set of remarkable (and rather under-appreciated) pieces that forge an idiosyncratic path—one that is notably different from his earlier work even within his late style.

1.3 Methodology

The term "macroharmony" already hints at the kind of large-scale analytical approach that is needed to adequately represent the way in which a listener interprets the presence of dominant scale areas in a piece of music. This paper uses visual representations in the form of various types of graphs to gauge the nature of these broader macroharmonic patterns; they essentially serve as compressed maps of the scalar structure in the textures and their juxtaposition, and in the analysis will be a starting point for the commentary.

This section shall present a conceptual overview of the types of graphical representations used in this paper, without detailed illustrations or discussions—though examples of most graph types will be presented for initial reference, using a section of Etude 15 that will be the subject of the next chapter. An in-depth explanation and dissection of each graph type, including the way in which they are created, the calculations involved and the analytical insights that can be derived from them, is the focus of Chap. 2.

The graphical representations in this analysis plot the progress of each etude in "rhythmic units" on the x-axis against one or more variables on the y-axis. A rhythmic unit (henceforth RU) is defined as each consecutive point of attack—which, given the aforementioned *perpetuum mobile* nature of virtually all of the passages of these etudes, yields a completely even, regular measure of time, creating a convenient rhythmic/graphical grid. There are only three short passages in the four etudes of Book 3 (one of which is illustrated in Fig. 1.3) where the proportions of this underlying grid seem to change. However, all of these are, in fact, written-out *ritardandi*, and therefore the progressively expanding note lengths do not affect the data.

Another feature of these etudes that proves convenient for this analysis is the fact that the duration of the vast majority of the notes corresponds to their RU, meaning that each note is processed as a single datapoint. (Namely, only the pitch or PC at

Fig. 1.3 Etude 16, RUs 159–175, score. © 2005 Schott Music, Mainz—Germany

Table 1.1 Graph categories and types	Category	Type
	P and PC representations	Piano roll
		Chromatic PC
		Co5 PC
		Co5 conical PC
		Spanned chromatic PC
		Spanned Co5 PC
	Correlation strength	Six-PC
		Seven-PC
		Eight-PC
		Spanned comparison
	Scalar force	Chromatic
		Tritone
		Hexatonic
		Octatonic
		Diatonic
		Whole-tone
		Spanned comparison

each RU is registered, rather than requiring another parameter for note length.) Only a handful of notes in these etudes last longer, and discussion later on in this chapter will address how these cases are dealt with in the analysis.

The graph types can be placed into three broad categories, as shown in Table 1.1. Pitch and PC representations constitute the first such category. The first graph type is a "piano roll" (Fig. 1.4), plotting the notes in each work in pitch space. This is the only graph type to do so, as the others are all situated in PC-space. Hence, the piano roll gives an insight into registral patterns that are not evident in the other visual representations, whether they are broader movements up and down the piano keyboard or the more local ordering of PCs in pitch space within a particular texture.

The next graph types convert the piano roll from P-space to PC-space. These are essentially bubble charts: a normal-size bubble marks one occurrence of a PC at a given RU, a slightly larger bubble two occurrences. (In these etudes, no more than

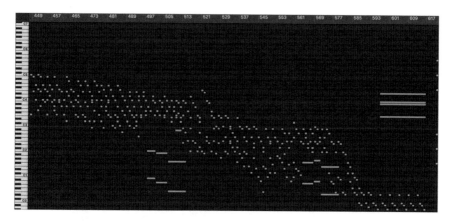

Fig. 1.4 Etude 15, RUs 448–621, piano roll (notes in P-space, C3 = middle C). P-space is used as an abbreviation for pitch-space

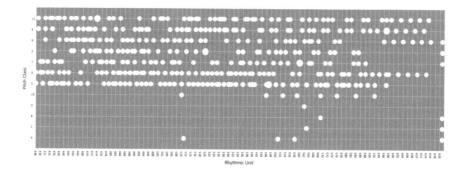

Fig. 1.5 Etude 15, RUs 448–621, Co5 PC graph (notes placed in PC space ordered by the Co5)

two pitches ever belong to the same PC in any one simultaneity.) There are two ways in which the PCs are placed on the y-axis: first in chromatic order, and then arranged in the circle-of-fifths (Fig. 1.5).[6] Hence, pitch space has been transformed into what this paper refers to as "chromatic" and "harmonic" PC space. The latter type of graph is considerably more important and revealing throughout this analysis, giving a first glimpse into the constitution of different macroharmonies as well as their juxtaposition and the manner in which the texture moves from one scale area to the next. This graph type is also more difficult to effectively represent on a Cartesian plane given the (evidently) circular nature of harmonic PC space—considered topologically, the graph is essentially a cylinder. Thus, the outer boundaries of the y-axis have to be arbitrarily determined, and the highest value is always one step "below" the lowest.

---

[6]Circle of fifths is hereafter abbreviated to "Co5."

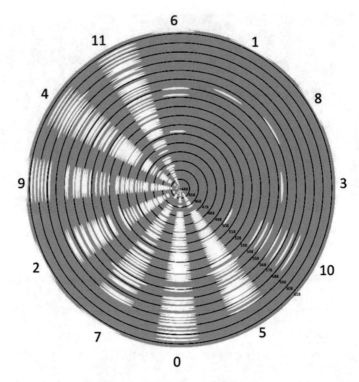

Fig. 1.6 Etude 15, RUs 448–621, conical Co5 PC graph (notes placed in PC space ordered by the Co5 using "polar coordinates": the y-axis is rendered as a circle, and the black rings mark the passage along the x axis)

The current investigation uses two workarounds to this problem. While the default arrangement of PCs on the y-axis places the white-note set at the top (with PC 11 at the upper end of the axis—in the same position as it is in the chromatic ordering) and the black notes below (with PC 6 marking the lower boundary), there are occasionally case-by-case re-orderings, depending on the nature of the etude or specific passage question; patterns are first gleaned from the default layout, which then inform the nature of the adjustments for the following Co5 PC graphs when necessary. The second solution is one that involves some graphical complexity (Fig. 1.6), where each axis operates on two dimensions: somewhat resembling the cross-section of an onion, the y-axis is rendered as a circle and gets progressively bigger for each passing RU, and the x-axis (and thus the progression of time) also moves outwards. This type of graph maintains the integrity of the shape of the Co5, and, while not straightforward to read, it provides an important complement to the rest of the PC representations.

So far, all representations described have taken the notes of each RU individually: they are plotted on the y-axis, without the resulting data for each RU being affected by what comes before or after. While this is useful for deriving some initial conclusions about patterns present in a texture, it is not reflective of a listener's

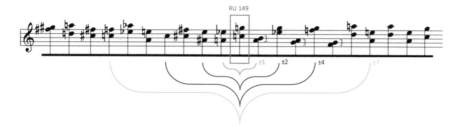

Fig. 1.7 Etude 18, RUs 139–159, marked score—RH only (±1, ±2, ±4 and ±7 spans around RU 149)

experience of the various scale areas of a passage, which comes from a summation of several consecutive RUs—rather than each simultaneity taken separately. As this book is a study of *macro*harmony, the focus needs to be on larger stretches of music rather than individual verticalities.

It is for this that the analysis uses "spans." A span is essentially how "zoomed out" a view we are taking for the variables present at a certain RU, expressed as an integer preceded by the ± symbol. The graphs so far all processed the pitch or PC data using a span of ±0, namely each RU individually. A span of ±1 would count the data from any RU as well as the one right before and right after as a single data point, in total a stretch of three RUs; ±2 would include two before and after, namely a stretch of five RUs, and so forth (demonstrated by Fig. 1.7).

Using the data from Fig. 1.7 as an illustration, the first graphical representations described above would simply take RU 149 individually; the PC "input" for this point is {0, 7}. However, with a ±1 span the input would be {0, 3, 7, 9, 9, 11}; for ±2, it would be {0, 3, 3, 4, 7, 7, 9, 9, 10, 11}, and so forth.

This facilitates a type of graphing that can show the strength of various variables as products of broader musical structures while at the same time sensitively tracking the changes in these variables from one RU to the next: not only are larger "chunks" of music investigated, but the "frame" of the chunks also move along with the progress of the piece, providing a highly dynamic view of large-scale processes.

In the graphical representations the spans can be used in two ways. The first is to take a fixed span and use it to compare different variables. This is essentially the process that has taken place so far: as mentioned above, the graphs described above were ±0 span representations, comparing the occurrences of different PCs—or, in the case of the piano roll, pitches. Thus, the final types of graphs in this first broad category takes the chromatic/Co5 graphs (in the default ordering) and applies a certain "representative" span (such as ±6) to them (Fig. 1.8). This renders the bubble charts considerably more intricate: instead of having two bubble sizes, the plotted points now vary considerably in magnitude given the accumulation of input, forming "bands" of constantly changing width along the horizontal course of each graphed PC. This type of visual representation helps to clarify certain trends perceived in the initial PC graphs: it facilitates the comparison of the relative presence of PCs using the varying size of their corresponding ribbons (rather than just from their frequency

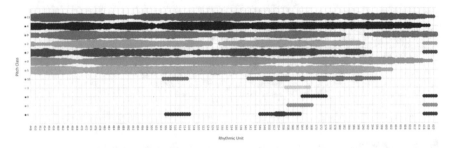

Fig. 1.8 Etude 15, RUs 448–621, Co5 PC graph, ±6 span (notes placed in PC space ordered chromatically, with a span of ±6 applied to each RU)

along the x-axis); this summation of data also makes these graphs a better reflection of how the rapid textures are perceived by the listener.

Given the quick speed in most passages of these etudes, the standard range for analysis is ±2 to ±10—any smaller is too short to be identified as a possible scale area, and any larger renders the resulting data too insensitive to the density of the processes. Spans within this range are also presented in the graphs, but they are limited to three in number to facilitate readability and prevent the overcrowding of data when multiple spans are shown together in one representation; the integers ±4, ±6 and ±8 are chosen for an even spread (which yields five data lines in total, including the minimum and maximum spans). The representative span is generally the halfway point between ±2 and ±10, namely ±6.

Whereas the first category of graphs focused on the dynamics of particular pitches or PCs, the second category departs from it entirely. Here, it is not the specific PCs that are of interest, but their broader statistical proportions within a certain passage. The graphs in the second category also largely focus on the second of the two ways that spans can be used graphically, alluded to above: the comparison of different spans within a single variable.

This second category focuses on an idea called "correlation strength." While the PC graphs provide an idea of which scale areas are dominant in a certain passage, they only roughly represent the extent to which the entire texture in question adheres to that collection of pitches. Namely, we may observe some outliers and make a guess as to the proportions of PCs within and outside of the macroharmony, but there is no readily available, precise figure for it. This is what the correlation strength seeks to do.

The correlation strength is defined as "the ratio of the number of notes that belong to the x most frequently-appearing PCs to the total number of notes." The x refers to the scale size (in terms of PC-set cardinality) being investigated. If, for example, the PC graphs seem to indicate a centering around a seven-note (perhaps diatonic) scale in a certain stretch, an x of 7 would be investigated. The number of notes in the passage that belong to the seven most frequently appearing PCs divided by the total number of notes in the passage will give the seven-PC set correlation strength (Fig. 1.9). This figure does not express *which* PCs those are, or what PC set they

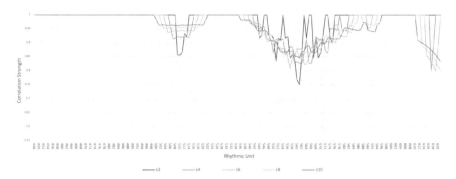

Fig. 1.9 Etude 15, RUs 448–621, seven-PC correlation strength (number of notes that belong to the seven most frequently appearing PCs/total number of notes)

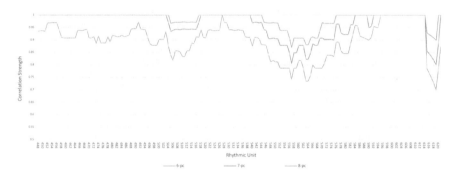

Fig. 1.10 Etude 15, RUs 448–621, correlation strength comparison (comparison across six-, seven- and eight-PC sets using ±6)

correlate to. However, in a passage where a particular dominant PC set is identified from the PC representation graphs described above, the correlation strength metric serves as a powerful reflection of the extent to which a listener will perceive the scale area as "clean" or "obscured" (or if they will perceive it at all), and tracking the correlation strength along the course of longer passages of music is invaluable in determining the structural implications of macroharmonic stability and instability.

Most of these graphs compare the correlation strengths of a number of different spans for a fixed PC-set size, allowing the comparison of different "zooms" to each other; this helps to clarify the temporal scale on which these procedures take place, for example to identify the "boundaries" of a macroharmonic area or the rate of a modulatory process. A final graph takes a representative span for comparison across different PC-set sizes (Fig. 1.10).

The final, third category of graphs is in a sense a combination of the first two, presenting data that reflects both the nature of pitch-class sets *and* the passages' correlation to them. This is effected by adapting the "Fourier Balances" (Fig. 1.11) devised by Ian Quinn in "General Equal Tempered Harmony" (2007), in turn

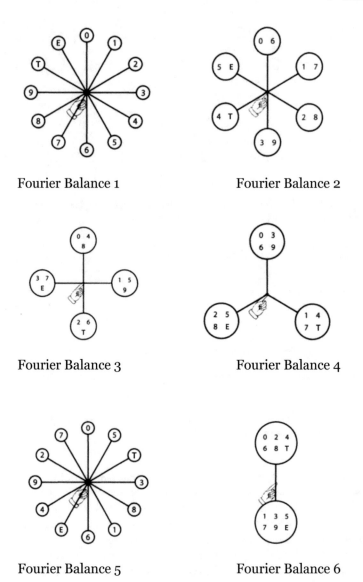

Fourier Balance 1

Fourier Balance 2

Fourier Balance 3

Fourier Balance 4

Fourier Balance 5

Fourier Balance 6

Fig. 1.11 Fourier Balances. These Fourier Balance diagrams have been adapted from Quinn (2007), and the subsequent visualizations of Fourier Balances are also all based on these six graphics

derived from David Lewin's "Fourier Properties" (Quinn 2007, 30 and Lewin 2001, 5–6).[7]

---

[7]For an extended discussion on the Fourier Transform, see Emmanuel Amiot's "Music Through Fourier Space" (2016).

Each balance places the 12 PCs on pans connected by panhandles at a central axis. The balance numbers correspond to Lewin's Fourier Properties (which they initially serve to illustrate), namely the property of a chord where it has the same number of notes in certain types of sets as it has in their complements. For example, a chord has Fourier Property 6 (also referred to as whole-tone-scale property) if it contains the same number of notes in one whole-tone set as it does in the other (Lewin 2001, 5). This would cause a perfect balance on Fourier Balance 6. If there were more notes belonging one whole-tone scale than the other, it would cause a tilt.

However, it is *precisely* that tilt that is useful for our purposes. A detailed discussion of how this yields examples of what Quinn calls harmonic *genera* can be found in his article (Quinn 2007, 30–45); here it is enough to imagine PCs causing a downward pushing force on the balances, and observe the types of complementary PC sets that, if operating individually (i.e. without their complement), cause a maximum tilt (Fig. 1.12).

As can be seen in Fig. 1.12, maximum imbalance is caused by PC sets that exert force on geometrically adjacent pans and occupy the widest possible span on the balance before a diametrically opposite PC pair is reached (which would cause a weakening of the force due to a "balancing out"). Hence, a geometric definition of a set that causes maximum imbalance would be a collection of PCs that occupy all adjacent pan handles *within* a 180° angle (this division is marked by the dotted blue line in Fig. 1.12). Also, it is worth noting that PC sets that occupy adjacent panhandles exert greater force than those that "skip" spots in between.

The resulting PC sets from each graph—considered in prime form, as they are transpositionally equivalent—yield mostly familiar scale types that are salient for the current analysis. For balance 6, it is the whole-tone scale; on balance 4, the octatonic, and balance 3 the hexatonic (namely of the hexatonic collection 014589; in this analysis the term hexatonic is only used for this type of six-PC scale). Balance 5 yields a set that is *almost* the diatonic scale—024579 (in prime form), six adjacent steps on the Co5, also known as the Guidonian hexachord. The seventh step, PC 6, causes a "balancing out" with zero. Hence the diatonic hexachord exerts a stronger force than the diatonic scale (which will prove to be a highly important consideration at certain points in this book) but in any case, balance 5 is representative of "diatonic force."

The first two balances are less familiar. The PC sets that exert maximum force on each balance can both be referred to using Babbitt's labels for all-combinatorial hexachords: A for balance 1 (012345) and D for balance 2 (012678). The former is here called "chromatic force," which is reflective of chromatic *clustering* rather than saturation (which would cause a balancing out of the force), and the latter "tritone force," given the three pairs of chromatically adjacent tritones that form the set.

These balances are most useful here not to simply plot PC sets on the balance for the sake of visual illustration, but to quantify the force applied: Quinn presents a system for measuring the total force of any PC set on a balance in *Lewins* (or Lw), where the force of any (single iteration of) one PC in isolation yields 1 Lw (Quinn 2007, 42–45). For the purposes of this analysis, as will be detailed further on in this investigation, a rather different method is used for the calculation of Lewins, but

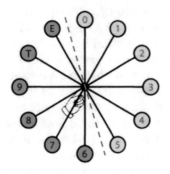

Fourier Balance 1 (Chromatic Force)

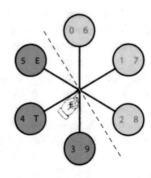

Fourier Balance 2 (Tritone Force)

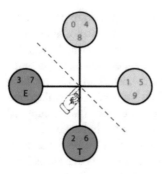

Fourier Balance 3 (Hexatonic Force)

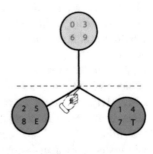

Fourier Balance 4 (Octatonic Force)

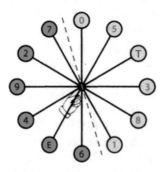

Fourier Balance 5 (Diatonic Force)

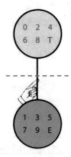

Fourier Balance 6 (Whole-Tone Force)

Fig. 1.12 Shaded Fourier Balances with PC set types that cause maximum tilt (green and red mark complementary sets; dotted blue line shows "tilt axis")

Fig. 1.13 Etude 15, RUs 448–621, tritone force graph (force applied to Fourier Balance 3, measured in Lewins—span comparison)

both approaches are based on the geometric properties of the forces drawn as vectors on a Cartesian plane.

Hence, these Lewins provide yet another metric for tracking macroharmonic changes, one that expresses both the strength and the type of scales (especially when comparing the results from different forces). Like the correlation strength, it cannot show which pitches or PCs exactly are prominent—for that we have the first category of graphs—but it gives an insight into the *patterns* of PCs that are prominent. The force graphs are also similar to those for correlation strength in the way they are presented. Initially, different spans are compared within one variable— namely, one type of force (such as "octatonic force," shown in Fig. 1.13)—before a final graph compares all forces using a representative span (Fig. 1.14).

All of the force graph types prove to be highly useful, even in passages that do not employ those particular scale types per se. While the utility of balances 3 through 6 is immediately evident—as they refer to more commonly used scale types—the patterns on balances 1 and 2 also often constitute a key consideration in the analysis, for example due to their reflection of what this paper calls "scalar distribution," namely, how evenly the PCs in a set are spread across the gamut of 12 PCs ordered chromatically. Generally speaking, the higher the Fourier Balance number, the higher the scalar distribution of the scale type that causes the maximum tilt, with the PC set for balance 1 (012345) being maximally uneven (as it is clustered within one half of the chromatic scale), while the whole-tone scale is maximally even. It is worth noting that evenness does not necessarily equal regularity. While the series of interval-classes 2-2-1-2-2-3 for the steps along the diatonic hexachord is considerably more even than that (1-3-1-3-1-3) in a hexatonic scale (with regards to the similarity in sizes of adjacent interval classes), the pattern itself is less regular in that it shows fewer degrees of rotational symmetry (one, compared to three for the hexatonic).[8]

The force graphs can also (directly or indirectly) reflect other processes, even those relating to interval choice and voice-leading—for example, the presence of

---

[8] Interval class is henceforth abbreviated to IC.

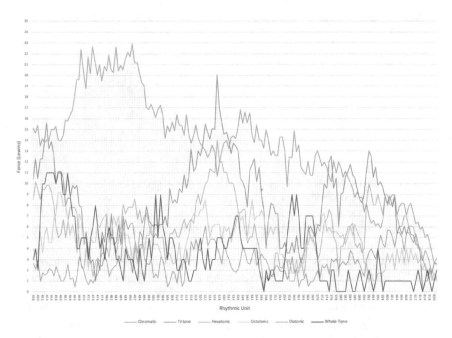

Fig. 1.14 Etude 15, RUs 448–621, force comparison (force applied to the six balances, using ±6 as a representative span)

perfect fourths or fifths moving in chromatic steps will register a strong tritone force. In the force comparison graphs, the manner in which the forces alternate and change (such as the order or rate at which they do so) also provides an invaluable window into the nature of modulatory procedures.

There is one other visual representation used in this book, one that is specific to the analysis of Etude 16, *Pour Irina*: a line chart that is used as a supplement to the other graphical representations. Given that the data input for this graph type is an interpretation of the modulations in the passages in question (rather than "raw" pitch or PC data) it is not included in this methodology and will instead be discussed separately in Chap. 3.

In addition to the above graphs used in this book, scrolling graph videos for each etude have also been created. These present the three most frequently used graph types—the Co5 PC, seven-PC correlation strength and ±6 force comparison graphs, aligned with a notation row above them—scrolling vertically with the progress of the music, synchronized with a software-produced recording. Links to the videos for each graph and a short discussion on how to interpret them can be found in Appendix A.

Taken all together, these graph types are something of a map or x-ray not only of the etudes themselves but also of the aural process, attempting to approach a visualization of how the listener may perceive the presence of and movement between different scale areas and the resulting dramatic narrative. However, it is also worth clarifying the features of a passage of music that influence a listener's perception of macroharmony that these graph types cannot—or will not—show.

As mentioned at the outset of this section, except for the piano roll, all of the graphs deal with PC space, rather than that of P-space, and hence fail to take into account registration, a feature that is in many cases central to our perception of tonality. Hence, it is it necessary to keep an eye on the score and the piano roll when interpreting data from the rest of the graphs.

One feature that theoretically could be reflected by the graphs are note lengths (namely, the duration of notes that last beyond the RU of their attack, of which there are only a handful in the etudes of Book 3) but this paper will not do so for a number of reasons, instead only recording the data at the point of attack. The primary consideration here is nature of the piano as an instrument—the notes decay in sound soon after they are struck.[9] As the volume diminishes in later RUs after the attack point, their inclusion in the data as being constantly present (hence having their attack repeated for each RU) would be a misrepresentation of the aural effect and a heavy-handed distortion of the data. (If this were for example an organ work, the analysis would take the opposite approach.) An attempt to instead make an estimate of how much of the note is still present with every passing RU would be largely doomed to imprecision, while also highly dependent on performance, acoustics and so forth. Another reason is that the actual lengths for many of these longer notes are not strictly defined in the score. (See, for example, the discussion of *White on White* in Chap. 2.) Hence, it is difficult even to determine until when the notes should in principle be sounding. That is not to say that note lengths are not considered in the analysis—they are a key feature that influence how we perceive the primary notes in a texture (and hence resulting scale area) and need to be taken into account, but simply cannot be adequately quantified numerically.

Another important aspect of the music that the graphs simply cannot reflect are dynamics. These visual representations in fact work much better for Ligeti's four final etudes than they would for most of his previous ones, given the absence of accented polyrhythmic patterns on various dynamic planes—but there are still occasional accents (with the rest of the notes in the corresponding textures adhering to one dynamic level), and loud notes are inevitably placed in the aural foreground, thus having oversized importance in our perception of tonality. It is for these reasons that the graphs are considered a starting point, an initial breakdown of the forces at play in an etude that will then, when placed in the context of the musical textures, yield insights into the structural and dramatic effects of macroharmonic juxtaposition in these works.

This approach to the graphs is reflected in the structure of the central analytical chapters that focus on each etude in turn. Following an initial presentation of the principal features of the etude in question and the main arguments of the chapter, each chapter presents a "graphical analysis" of the work—a representation of the complete musical material in each of the graph types—that serves as a visual

---

[9]This occurs faster in the higher registers of the keyboard than in the lower ones.

summary and representation of the main macroharmonic processes in the piece. This is then followed by a detailed commentary that follows the structure of the work, discussing the musical material and the way in which the processes within are reflected and illuminated by the graphical representations.

While the core analysis in each chapter largely follows the structure of each etude, one chapter is arranged somewhat differently: Chap. 2, on Etude 15. This is due to the fact that there is only roughly one page of music (the last out of five) that can be thoroughly analyzed for macroharmonic change, as until then the rest of the work merely employs white notes—which, though in itself interesting, renders most of the variables that form the analytic focus of this paper largely static, given the rigid and unchanging nature of the scale area. The etude's conclusion on the other hand, with its light dusting of "foreign" notes, serves as a perfect vehicle to demonstrate the sensitivity of the graphs' metrics in responding to macroharmonic changes.[10]

This passage is used as a kind of preliminary study, a "test run" to present these analytic tools in action. While the other chapters are centered around the discussion of musical material and use the graphs for illustration and elucidation, Chap. 2 will do the inverse. After a brief discussion of the main features of *White on White*, the chapter will present each graph type in the foreground—using the conclusion of the etude as the "source data"—discussing in detail how each graph is formed, the manner in which it presents data and the types of patterns that can be observed, while also drawing a number of conclusions about the musical material from the graphs as an illustration of how they are interpreted in the rest of the analysis.

To conclude this chapter it is worth discussing the analytical scope of this paper and the relationship of these techniques to previous research. Given the emphasis on macroharmony and especially the relationship between scale areas and large-scale compositional processes in rapid, highly dense textures, the main focus of this book is *not* in the constitution and harmonic profile of individual simultaneities (and the voice-leading or harmonic motion between them) and hence diverges in both methodology and the nature of the material investigated from other analyses that use similar tools, such as Fourier Balances. The sheer scale of the data, whether in the number of temporal moments investigated (namely the thousands of RUs) or the density of data for each one (given spans that, in the densest textures at the maximum ± 10, can yield sets of up to 100 objects for each RU) requires a different and unique approach in order to present an adequate broader summation, a bird's-eye view— one for which there is no truly similar analytical precedent. (To produce the graphs in this book, the software used had to in fact execute over 30 million calculations.)

To illustrate one aspect of this difference using the example above, it is worth considering for example the approach of Quinn (2007) and Howard (2014) in two analyses that also employ Fourier Balances: they discuss the effect of the various PC sets on the balances without invoking the question of the cardinality of the individual PCs themselves if investigating a particular passage of music (namely, if more than one note in the set analyzed belongs to a certain PC), given that Quinn is

---

[10]The term foreign denotes a PC that is not part of the macroharmony.

investigating harmonic space in a generally abstract sense (namely, not focused on a specific musical work), and Howard does not consider PC doublings as salient in the analysis of the Feldman piece (hence only dealing with the "chord quality" of each simultaneity). In this investigation, however, the frequency of the appearance of particular PCs within a texture investigated is in fact central to the understanding of the macroharmony—for the listener, and in turn the analysis.

This is not to say that the question of harmonic quality is not occasionally discussed in this book; in certain passages it proves important to our understanding of broader macroharmonic or structural patterns, and for these the work of analysts such as Shaffer (2011) have served as a key influence in understanding the nature of harmonic processes in Ligeti's late work—more so than Drott (2003) and Searby (1997) and Ligeti et al. (1983) for example, who have made a case for the reductive "surface sonority" theory of triadic harmony in Ligeti's late work, as mentioned earlier in this chapter.

This discussion would not be complete without addressing the work this paper has the most obvious connection to—Tymoczko's "Geometry of Music" (2011). As mentioned earlier in this chapter, here the claim is made that a certain passage may embody a particular macroharmony—a certain scale type, for example—even if not all of the notes in the passage adhere to it, namely that the macroharmony is fuzzy (thus departing from Tymoczko's definition of the term). The markers for correlation strength or scalar force are used not only to measure and quantify the extent of that fuzziness, but to show how the varying correlation to certain scale areas is in itself a compositional tool to affect a listener's perception of stability and instability, and hence a means to craft dramatic narratives. Therefore, the "macro"-ness of the harmony as understood here is in its longer-range presence from a horizontal perspective, rather than its all-encompassing nature vertically. In fact, some of Tymoczko's discussions of what he defines as features independent of macroharmony—such as pitch centricity—prove more salient to the analysis here, as they are key features that influence the way in which a listener might experience being "anchored" around a certain set of PCs or scale area, for example.

In addition, although Tymoczko uses a number of graphing techniques in analyzing macroharmony, they all deal with how a piece of music uses it on a global scale, rather than tracking the progress of the associated variables across the course of the work from one unit of time to the next—which is precisely what this paper sets out to do, in order to somehow simulate (or at least reflect) the experience of listening to the work, and thus resulting in a very different analytical approach.

Chapter 2
Alternating Diatonic Qualities in Etude 15, *White on White*

As mentioned in the introduction, the perhaps defining feature of the etudes of Book 3 is the restricting of the gamut of PCs used at any one time, as opposed to the freer chromatic saturation of his previous piano etudes. There is perhaps no more radical example of this than Etude 15, *White on White*, which for the majority of its duration does not stray at all from the white-note set—and the few black notes that do appear towards the end of the piece come across as specks of pepper in a sea of salt.

That is not to say that this extreme diatonicity is necessarily "tonal."[1] On the contrary, for a large portion of the work Ligeti goes out of his way to avoid the implication of tonality, embodied best by the absence of triads, which in such an exclusively diatonic texture requires a concerted effort to achieve. As will be shown in the analysis, the absence (and eventual presence) of triads is a highly important feature of the structure and dramatic narrative of the etude.

White on White is in two primary sections. The first is a leisurely *Andante con tenerezza* in half-notes where the LH follows the RH in canon by one RU, one octave below (Fig. 2.1). The *Andante* has three subsections, all of which are harmonizations of a melody of 40 RUs (5 bars of 16 RUs each)—the first and third of these are identical, creating a kind of ABA structure. This kind of literal repetition—unprecedented in Ligeti's etudes—is another clear sign of the new (or in certain respects, "old") direction the composer is taking in Book 3.

The *Andante* is followed by an energetic *Vivacissimo con brio* in running eighth-notes that is in two main parts, the first (starting at RU 121) brash and dry, and the second (from RU 448 onwards) hushed and luminous—which is where the "guest" notes begin to appear. The difference in sonority between the two sections of the *Vivacissimo* is not merely due to the dynamics (*fortissimo* as opposed to *pianissimo* progressing to *pianissississimo*) and the pedaling (*quazi senza Ped.* followed by sustain pedal marks and *una corda ad lib.* in the conclusion). It is also due to the

---

[1] For a discussion of the tonal implications of the diatonic set see Browne (1981) and Gauldin (1983), as well as Clough (1979), Egmond and Butler (1997), and Santa (2000).

© The Author(s), under exclusive license to Springer Nature Switzerland AG 2022
N. Namoradze, *Ligeti's Macroharmonies*, Computational Music Science,
https://doi.org/10.1007/978-3-030-85694-6_2

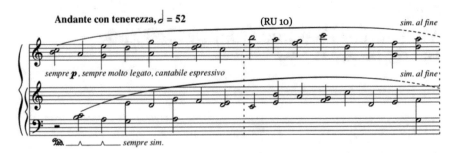

Fig. 2.1 Etude 15, RUs 1–16, score. © 2005 Schott Music, Mainz—Germany

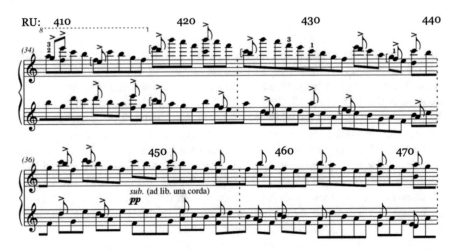

Fig. 2.2 Etude 15, RUs 409–472, score. © 2005 Schott Music, Mainz—Germany

intervallic relationships of the simultaneities. The initial *Andante* contains chords that are composed primarily of stacked seconds or fourths (in PC space, not necessarily pitch). The *Vivacissimo* seems to sift this into two groups: the first section is composed entirely of diatonic clusters, while the conclusion is dominated by stacks of fourths (again, both in PC space). Figure 2.2 shows the boundary between the two sections at RU 448, on either side of which is the representative texture for each.

The conclusion of *White on White* (RUs 448–621) is the only part of the etude where the white-note set is "breached," and thus forms the focus of this chapter, which makes four primary claims: the appearance of black notes in this section does not alter the general macroharmony but makes it fuzzier, "greying" the white-note set, so to speak; these appearances follow a clear pattern defined by the Co5; the change in intervallic composition of the texture at RU 448 is reflective of the process that facilitates this movement; and finally, that this process articulates a larger conflict in the work between what this paper calls "linear" and "harmonic" diatonicity.

The shift in bar 36 can be seen not as a movement away from clusters, but rather to a different kind: from simultaneities constituting adjacent steps in diatonic PC space, to those in "harmonic" PC space (i.e. the Co5). Namely, the first part of the *Vivacissimo* uses consecutive PCs along the C-D-E-F-G-A-B ladder, where one step beyond the scale in either direction still yields a continuation of this set: a "linear" diatonicity where the system is full or "closed." However, as soon as the conclusion reorders these steps as F-C-G-D-A-E-B, the system is open: if a harmonic cluster (namely a PC set constituting adjacent steps on the Co5) moves too close to the edge and oversteps this frame we instead get Bb, F# and so forth—which is exactly what happens.

While the graphical analysis for the chapter focuses on this final passage, some examples will show patterns across the entire work to demonstrate the relationship of the conclusion to previous textures. The conclusion of the chapter will place the procedures at the end of the piece in the larger context of the whole etude, in particular to develop the fourth claim made above.

It is also worth noting that the above use of the term "greying" does not derive merely from the introduction of black notes in the white-note set, but also from the image of the Suprematist painting *White on White* by Kazimir Malevich (1879–1935), from which this etude possibly takes its name (Fig. 2.3). Though we cannot know if this is a deliberate parallel on the part of the composer, one could say that Ligeti's Etude in a certain sense musically reflects an idea that the painting presents, that of different shades of white: one that is colder and harsh (i.e. the fully white-note passages until RU 448) and the other softer, warm and less pure (namely the final greying at the conclusion). This macroharmonic shading is an illustration of the way in which the term "macroharmony" is used here.[2] As mentioned in the methodology, this is a musical process that is particularly effective for displaying the sensitivity of the graphs and how they reflect features of a texture.

As discussed in the Introduction, the graphical analyses take the foreground in this chapter: various types of graphs are presented in turn, and though the focus of the discussion is on the manner in which they are produced and the ways to interpret them, it also addresses a number of conclusions that can be derived from these representations. Given that almost all of the graphs use the final section of *White on White* (namely, RUs 448–621) as the data source, the entire passage is presented in Fig. 2.4 for reference (though a micronotation row tracking the data will accompany most of the graphs).

Despite the sudden change in sonority at the *pianissimo*, this boundary is considered a "soft" cut given the fluidity of the rhythm and textural motion. Hence, it is treated like the "part" divisions described in the Introduction: the data from the few RUs leading up to the dynamic change are still included in the spans that entail them.

---

[2] Namely, as opposed to Tymoczko's use—see the discussion on page 21.

Fig. 2.3 Malevich, Kazimir (1878–1935): Suprematist Composition: White on White, 1918, New York, Museum of Modern Art (MoMA). Oil on canvas, 31 1/4 × 31 1/4′ (79.4 × 79.4 cm). Acquisition confirmed in 1999 by agreement with the Estate of Kazimir Malevich and made possible with funds from the Mrs. John Hay Whitney Bequest (by exchange). 817.1935 © 2021. DIGITAL IMAGE © July 16, 2021, The Museum of Modern Art/Scala, Florence

2.1 Pitch and Pitch Class Representations

There are six types of graphs in this category. The most basic of them is the piano roll: a visual representation of the notes of a piece in pitch space. The y-axis uses a keyboard for orientation with MIDI pitch labeling where C3 is middle C.[3]

Despite the focus in this chapter on the conclusion of *White on White*, it is worth first taking a look at the piano roll for the entire etude (Fig. 2.5), as this is the only

---

[3]Many of the graphs are shown in landscape orientation for better readability, where these axes are thus flipped at 90°.

Fig. 2.4 Etude 15, RUs 448–621, score. © 2005 Schott Music, Mainz—Germany

Fig. 2.5 Etude 15, RUs
1–621, piano roll (notes in
P-space, C3 = middle C)

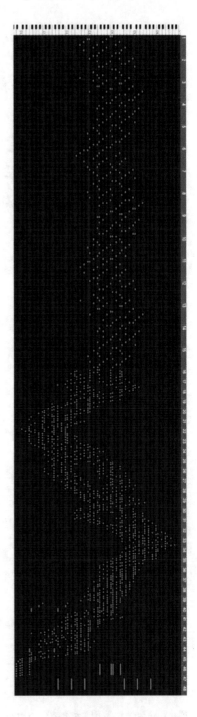

graph that displays significant variance over the course of the rest of the work (given that it explores pitch rather than PC space). *White on White* is also the only etude in Book 3 to use bar divisions (though they are dotted barlines, merely for orientation), so measure numbers have been used in this particular graph for navigation—bars 1–15 are the *Andante*, and bars 16–48 the *Vivacissimo*, with the conclusion beginning in the middle of bar 36. Marking the tempo change is a slightly more substantial vertical line, relative to the other gridlines; also, the measures in *Vivacissimo* are shorter than in the *Andante* as they contain half the total length of RUs (16 eighthnotes, rather than 8 half-notes).

The graph is a reflection of how the different sections use the keyboard's registers: the *Andante* is relatively static in the middle of the range, while the *Vivace* makes a zigzagging shape, first plunging to the bottom, then working its way up to the top of the treble, and then steadily working its way back down again—before a final chord that frames the central register the *Andante* occupied. It also clearly shows the extent to which the two hands largely move together throughout the etude, even in the *Vivacissimo* despite the section not being in canon. It is only in bars 25–27 that there is something of a gap between the two hands, where the RH ascends faster than the LH which hesitates for a while before following suit—a pattern that is demonstrated by the score (Fig. 2.6) but considerably better illustrated by the piano roll.

Figure 2.7 presents more a "zoomed in" piano roll that focuses on the conclusion (now using RUs on the x-axis instead of bar numbers for greater precision). It also includes a micronotation row above for orientation and comparison, as will the rest of the graphs for sufficiently short passages. (For landscape-orientation graphs, any more than 200 RUs requires scaling that renders the notes miniscule beyond legibility; the notation rows for those in portrait orientation are no longer than 50 RUs in length.) Wherever multiple graphs showing the same passage of music are superimposed for comparison, a single notation row is placed above them for reference. In order to ensure clarity and readability, such notation rows are devoid of dynamics, extra stems for voice assignations, or expression marks.

There is a discrepancy in the length of the last chord between the piano roll in Fig. 2.7 and the one in Fig. 2.5. This is because the last chord is a single, final RU, irrespective of its acoustic length (which is nevertheless maintained in the notation row in Fig. 2.7) and hence functions as a single data point. As mentioned in the introduction, the same will go for the notes that last beyond the RU they appear at— their data will only be recorded at the point of attack.

The next graph types are similar in construction to the piano roll but move from pitch space to PC space. One arranges the PCs on the y-axis in a chromatic order (Fig. 2.8a), the other by the Co5 (Fig. 2.8b)—the "harmonic" order. As mentioned in the introduction, for the Co5 arrangement the default is to have the white-note set above the black notes (namely, starting at 11 and moving down to 6); however, an additional Co5 PC graph may be added with a modified PC order in order to

Fig. 2.6 Etude 15, bars 24–29, score. © 2005 Schott Music, Mainz—Germany

demonstrate procedures unique to the particular work. In this etude, for instance, Fig. 2.8b shows a very clear pattern of introducing PCs foreign to the white-note set: a deliberate "widening." The first two to appear—almost simultaneously—are PC 10 (Bb) at RU 63 and PC 6 (F#) at RU 64, the first steps up and down the Co5. The next pair, again almost concurrently, are PC 3 (D#) at RU 115 and PC 1 (C#) at RU 116. This just leaves PC 8, duly entering as Ab at 122. (It is worth noting that there is no discernible pattern in the "chromatic" ordering, namely Fig. 2.8a; this procedure is clearly one governed by the Co5.)

Hence, a slightly different and more representative visual perspective on this widening of the gamut can be gained from a "reordered" Co5 PC graph, as shown in Fig. 2.9. The two examples also place the final chord—a pair of superimposed perfect fifths—in a slightly different light.

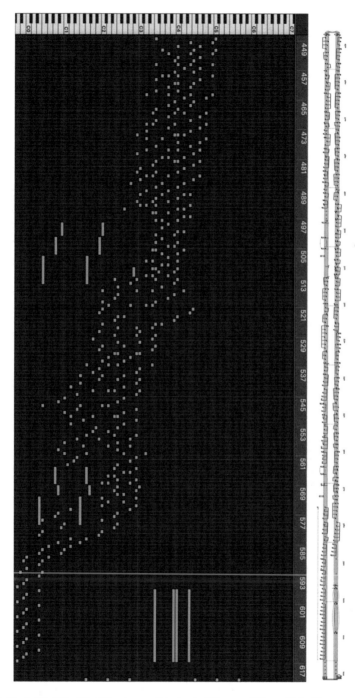

Fig. 2.7 Etude 15, RUs 448–621, piano roll (notes in PC space, C3 = middle C)

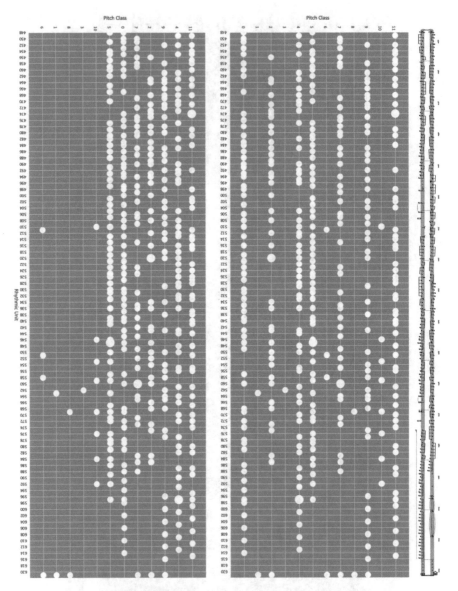

Fig. 2.8 Etude 15, RUs 448-621, PC graphs. (**a**) (top): chromatic (notes placed in PC space ordered chromatically). (**b**) (bottom): Co5 (notes placed in PC space ordered by the Co5)

It is clear from both graphs that the two portions of this final chord each symbolize the two opposing forces at play in this final section, namely the white-note set that is "grayed" by the encroaching black notes. The G-D-A chord in the LH is in the center of the white-note set's span along the Co5; Fig. 2.8b plots the F#-C#-G# far away, at the bottom of the y axis. This presents an interesting perspective,

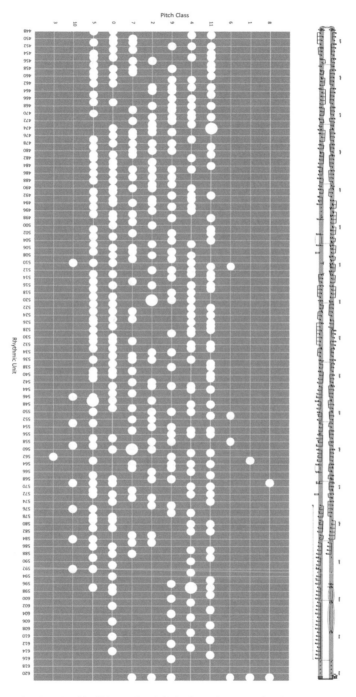

Fig. 2.9 Etude 15, RUs 448–621, reordered Co5 PC graph (notes placed in PC space ordered by the Co5, but with PC 8 instead of 11 as the upper value on the y-axis)

notwithstanding the inevitable linear distortion of the circular axis inherent in this type of visual representation: despite the almost simultaneous nature of the widening, the PCs just below the white-note set always come before those above (10 before 6, 3 before 1), hence the upper sharp area is indeed somehow "further" away than the flat area immediately below.

However, the "reordered" graph in Fig. 2.9 reveals the F#-C#-G# as not only as a counterpart to G-D-A by occupying the Co5 space above the white-note set, but also as an *extension* of the lower chord in both pitch *and* harmonic PC space: a glance at the piano roll (Figs. 2.5 and 2.7) reveals the (proportionally) exact same spacing of this chord. Namely, in both pitch space and harmonic PC space there are two "missing" steps in the middle of the chord (Fig. 2.10). This confluence of pitch and harmonic PC space in the final chord is reflective of the entire final section of *White on White*, where fifths take on a central role both as an interval in the verticalities and as a mode of modulation: the perfectly ordered, stepwise expansion from the white note set seems to show that the *Vivacissimo* did not simply switch from clusters to more spaced-out chords in bar 36, but rather moved from one type of cluster—in diatonic PC space—to another, in harmonic PC space.

However, the PC graphs also show that the white note set remains the principal, "central" scale—the black notes are nowhere near the rest in frequency of appearance and are in both the positional and cardinal periphery. Hence there is not a modulation of scale per se, but rather of the correlation to it—something that will be best illustrated by the correlation strength graphs in the following section.

There is yet another type of variety *within* the white-note set that cannot be illustrated by the PC graphs alone, or the correlation strength graphs for that matter—and that is the "harmonic quality" of the different passages that adhere to the white-note set. A quick glance at the Co5 PC graph for the entire etude (Fig. 2.11) shows a static pattern up until the widening that starts at RU 510. However, change in intervallic constitution at the "switch" in the *Vivacissimo* in bar 36 causes a palpable change in the quality of the macroharmony to the listener. To illustrate this, the force graphs—the third category of the representations—will be needed.

As mentioned above, no amount of reordering of PCs along the y axis in the "linear" Co5 PC graphs can truly reflect the way in which the texture moves *around*

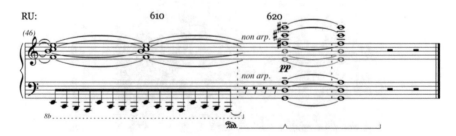

Fig. 2.10 Etude 15, RUs 601–621, marked score (final chord with "missing" steps—in orange—in both pitch and harmonic PC space)

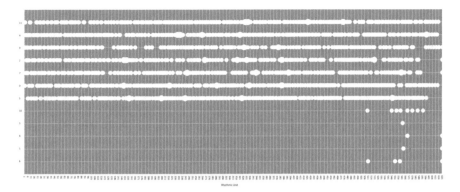

Fig. 2.11 Etude 15, RUs 1–621, Co5 PC graph (notes placed in PC space ordered by the Co5)

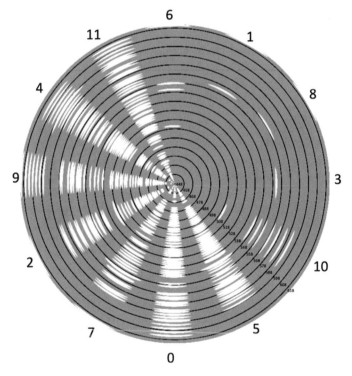

Fig. 2.12 Etude 15, RUs 448–621, conical Co5 PC graph (notes placed in PC space ordered by the Co5 using "polar coordinates": the y-axis is rendered as a circle, and the black rings mark the passage along the x axis)

the Co5, rather than just up and down it. The next graph type succeeds in representing the circular nature of the motion—the "conical" representation (Fig. 2.12).

This unorthodox representation is best understood if one imagines observing a cone from above, creating a two-dimensional view of a three-dimensional shape. The illustration in Fig. 2.13 illustrates this transformation.

Each "rung" of the cone is one RU (in this example, the black rings mark every tenth RU), with the progression of time moving down the cone. Hence, in the two-dimensional bird's-eye view, the x-axis begins in the center with the smallest circle and moves outwards. Each rung or circle is a manifestation of the y-axis, and the positions of the PCs are marked outside the cone. While requiring some familiarization to read, this method of representation is a powerful tool to track harmonic motion of passages around the Co5 along the course of a work. In this case, it shows how the boundaries of the white-note set in harmonic space—namely PCs 11 and 5—end up slowly getting "bridged" by the black notes, each new appearance of which helps to close the gap, so to speak. The switch from diatonic PC space to harmonic PC space at bar 36 has meant that a set that previously "filled" the system is now only partly covers it—and this incomplete coverage is best illustrated by the conical graph with the heavy "weighting" on one side of the circle. Indeed, it is as if this sudden "vacuum" is what induces the set to slowly expand outward, like areas of high and low pressure that begin to equalize.

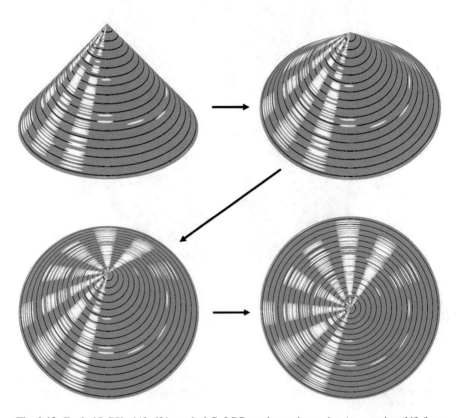

Fig. 2.13 Etude 15, RUs 448–621, conical Co5 PC graph, z-axis rotation (perspective shift from three to two dimensions)

The final two types of graphs in this category (Fig. 2.14a and b) are something of an intermediate step between the previous and following types of representation in that they are still PC graphs but now reflect the "span" concept that will become central to the rest of the graphs; they are simply PC graphs (chromatic and Co5, in turn) with a particular span applied to them. While they largely display the same patterns as the previous PC graphs, they serve to more closely simulate a listener's experience of dominant PC sets, where a note does not simply disappear from our consciousness as soon as it is gone but rather "lingers" in what we might recognize as the macroharmony. Thus, a cross-section of a "spanned" PC graph will give not only what PCs have an attack point at that particular RU, but what kind of general set is present in the area; they are essentially a "shaded in" version of the previous PC graphs. The span used here is ±6—the midpoint in the range of ±2 to ±10 used for *White on White*.

One of the effects of the spanning process is that it exaggerates certain patterns, making them especially evident. For example, both Fig. 2.14a and b show that of the three white notes in the final chord of the etude, only the A (PC 9) is active in the passage leading up to it (from RU 588)—G and D drop out. Not only are there no attack points for these two PCs in this stretch, but towards the end of bar 45 (at RU 597) there is a 045E chord in the RH that is held through the rest of the eighth-note texture (the duration of which is not picked up by the graphs), which also omits these two PCs (Fig. 2.15).

Figure 2.14b shows that the increasing (albeit brief) prominence of the Bb seems to temporarily pull out the white note set's center of gravity (as PCs 2 and 7 are two "central" PCs in the white note set on the Co5), a destabilization soon rectified by the final chord that is in fact grounded on these very two PCs. It is also worth noting that the final chord does not include any of the PCs of the preceding held RH chord—the sum of the two almost complete the total PC gamut (namely all chromatic PCs excluding 3 and 10).

Another purpose of this graph type is to present a look into how the "source data" is transformed by the spanning process—a key consideration in interpreting the rest of the representations. In particular, it shows how spans end up "anticipating" future events (as a span at a particular RU will be including data from future moments it encompasses); therefore, it is important to keep in mind that processes that appear in the spanned graphs usually only become evident to the listener a few RUs later.

2.2 Correlation Strength

The definition for correlation strength ("the ratio of the number of notes that belong to the x most frequently appearing PCs to the total number of notes") can be expressed in the following formula for a set of notes S, where S_n is the subset for the nth mode and the correlation strength c in a PC set of x PCs:[4]

---

[4]Letters for sets with vertical bars around them (i.e. $|S|$ and $|S_n|$) represent the cardinality of those sets/subsets.

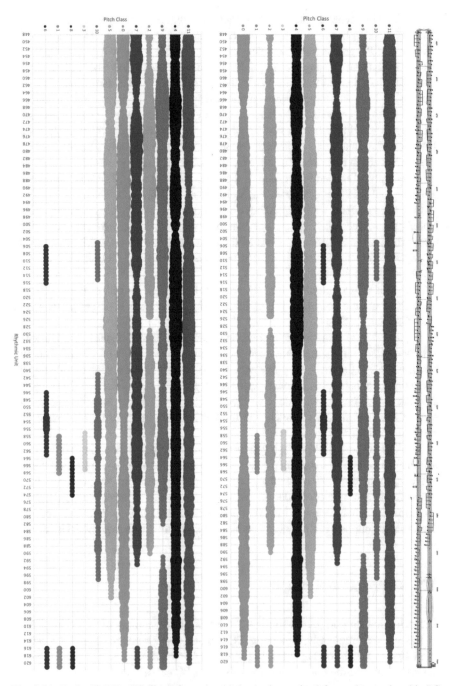

Fig. 2.14 Etude 15, RUs 448-621, PC graphs. (**a**) (top): chromatic, ±6 span (notes placed in PC space ordered chromatically, with a span of ±6 applied to each RU). (**b**) (bottom): Co5, ±6 span (notes placed in PC space ordered by the Co5, with a span of ±6 applied to each RU)

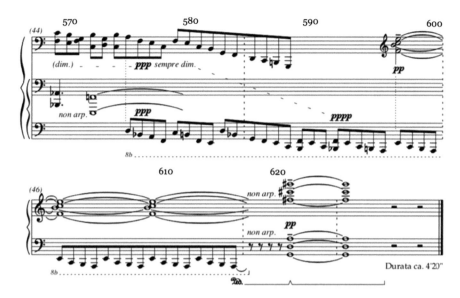

Fig. 2.15 Etude 15, RUs 569–621, score

$$c(S) = \frac{\sum_{n=1}^{x} | S_n |}{| S |}$$

For x, the paper looks at three PC set sizes: 6, 7 and 8. This is due to the fact that the scale types that are "represented" by the six Fourier Balances all contain 6–8 PCs, so the correlation strength is used in tandem with the force graphs to better illustrate the extent to which notes in a passage "belong" to a certain macroharmony or scale type.

For example, when looking at the correlation strength for 8-PC sets (when investigating octatonic scale areas for instance), if all notes in a certain stretch of music belong to the eight most frequently appearing PCs, then the correlation strength will be full, namely 1. However, if there are 25 notes in total but 23 of them belong to the 8 most frequently appearing PCs, the correlation strength will be 0.92.

The nth mode is here defined as the nth most frequently appearing PC in a set. To illustrate, the set of notes for ± 4 at RU 462 of *White on White* (namely all notes of RUs 458–466) is the following:

$S = \{0, 0, 0, 2, 2, 4, 4, 4, 5, 5, 5, 5, 5, 7, 9, 9, 9, 9, 11, 11, 11\}$

$|S|$ is 21. Here, the most commonly occurring PC is 5 which appears five times (namely the first mode, or S_1), the second most commonly occurring PC (S_2) is 9 which appears four times, and so forth. The subsets S_n can be listed as follows:

$S_1 = \{5, 5, 5, 5, 5\}$
$S_2 = \{9, 9, 9, 9\}$

$S_3 = \{4, 4, 4\}$
$S_4 = \{11, 11, 11\}$
$S_5 = \{0, 0, 0\}$
$S_6 = \{2, 2\}$
$S_7 = \{7\}$

Sets may be equal in cardinality, hence $S_1 \geq S_2 \geq S_3 \ldots \geq S_{12}$. However, the ordering (or ranking) of PC sets of equal cardinalities does not matter—in fact, the nature of the items in each set has no effect on the calculation, it is only the number of items in each set that is used.

If x is 6, the formula gives a $c(S)$ of 0.952, namely 20/21, as only one note out of the 21 did not belong to the 6 most frequently appearing PCs. When x is 7 or 8, the $c(S)$ here is 1.

There are four types of correlation strength graphs—one each for the 6, 7 and 8-PC sets, plus a final comparison graph across the three PC set sizes, using the representative (in this case ±6) span. These are shown in Figs. 2.16 and 2.17.

The y-axes for the graphs do not begin at 0, but rather at the minimum possible values for the correlation strengths: 0.5 for 6-PC sets, 0.583 for 7-PC sets, and 0.667 for 8-PC sets (rounded to 3 decimal places). These minimum values are reached when $S_1 = S_2 = S_3 \ldots = S_{12}$, namely when there is total evenness in the frequencies of all PCs in a passage (which never actually occurs in these etudes—but would happen in selections of 12-tone works, for example).

The correlation strength shows a pattern that could to a certain extent already be predicted from the previous discussion on the PC set graphs, with a first dip centered around RU 510–511 (where the initial Bb and F# appear), a second larger dip for the more numerous chromatic diversions at RU 546–592 (which registers a bit more broadly—namely from a bit sooner to little later—given the spanning) and a final weakening at the concluding chord.

Figure 2.16b is the most useful for *White on White*, given that the dominant macroharmony throughout is a 7-PC set (namely the diatonic white-note set). The plateaus of full correlation strength (i.e. 1) show the unsullied stretches of this scale area. The first dip is almost completely symmetrical given the nearly simultaneous breaching of the main 7-PC set—save for ±2 which displays a more erratic pattern throughout, as it is often not wide enough to encompass sufficient iterations of each white PC for them to clearly register as more frequent than the black PCs. Hence, one quite literally needs to zoom out a bit to get the big picture.

The shape of the large dip also is unusually symmetrical, a reflection of the acceleration-deceleration of the PC introductions in the passage: there are first three single black notes at RUs 546, 551 and 553, then three pairs (the first two on consecutive RUs 558–559 and 562–563, the last pair simultaneously on RU 570) and then again three singles on RUs 576, 584 and 592 (see Fig. 2.18). This sensitivity of the correlation strength marker to the modulation rate present in a texture makes it an important tool in analyzing different kinds of motion between scale areas—though in the case of this passage of *White on White*, it is not a

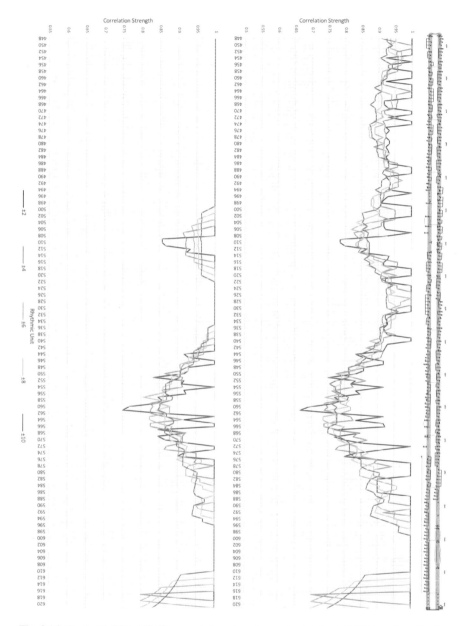

Fig. 2.16 Etude 15, RUs 448-621, correlation strength graphs. (**a**) (top): 6-PC (number of notes that belong to the 6 most frequently appearing PCs/total number of notes). (**b**) (bottom): 7-PC (number of notes that belong to the 7 most frequently appearing PCs/total number of notes)

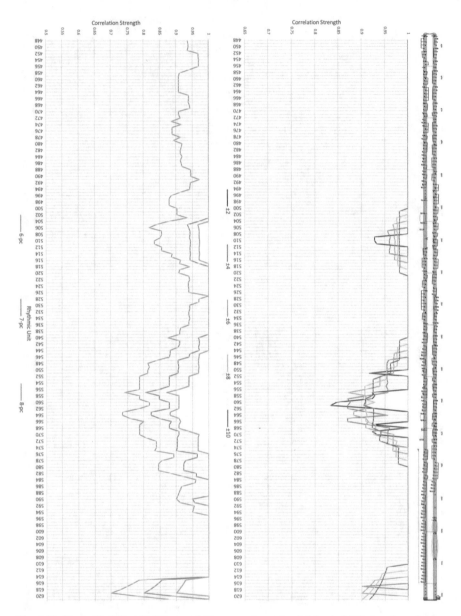

Fig. 2.17 Etude 15, RUs 448-621, correlation strength graphs. (**a**) (top): 8-PC (number of notes that belong to the 8 most frequently appearing PCs/total number of notes). (**b**) (bottom): comparison (comparison across 6-, 7- and 8-PC sets using ±6)

Fig. 2.18 Etude 15, RUs 537–621, score. © 2005 Schott Music, Mainz—Germany

modulation but rather a brief obscuring and subsequent clarifying of the same macroharmony.

This idea of clarifying or clearing of the air is reflected in an interesting manner by the 6-PC set graph (Fig. 2.16a). Before RU 594, the spans wider than ±2 never reach a correlation strength of 1 (for anything longer than an occasional, momentary spike), showing that the seven PCs of the white note set are always present in the texture—but upon the final LH ostinato (RUs 597–616) the correlation is total. It is as if the return from the moments of chromatic diversion to the white-note set overshoots its destination and ends up with a set even narrower than the original 7-PC set. This passage seems to clean the listener's aural palate, creating a blank (or "white") slate against which the final chord stands in sharper relief; as mentioned above, two of the three white notes in that final chord are actually missing from this ostinato (and the RH chord placed above it), therefore even those come as something of a novelty (see the discussions accompanying Figs. 2.10 and 2.15). In any case, the reversing of the modulation is an example of the way in which the white-note set exerts a gravitational force throughout—any strays from the path are promptly corrected and our sense of anchoring remains in place.

2.3   Scalar Force

The force graphs constitute the final category of graphical representations. Below is an overview of the mathematical principles used to derive the force metric (in "Lewins") before the graphs themselves are presented and discussed.

In order to calculate the force f that the set of notes S exerts on balance b, where the function $c_p(S)$ counts the number of instances of PC p in set S, one can use the following formula:

$$f(S) = \sqrt{\left(\sum_{p=0}^{11} c_p(S) \sin\left(30°pb\right)\right)^2 + \left(\sum_{p=0}^{11} c_p(S) \cos\left(30°pb\right)\right)^2}$$

This can also be expressed in radians instead of degrees:[5]

$$f_b(S) = \sqrt{\left(\sum_{p=0}^{11} c_p(S) \sin\left(\frac{\pi}{6}pb\right)\right)^2 + \left(\sum_{p=0}^{11} c_p(S) \cos\left(\frac{\pi}{6}pb\right)\right)^2}$$

This formula can be broken down into smaller parts. The demonstration here will use the same set S as in the correlation strength discussion above, namely the ±4 span at RU 462 (i.e. RUs 458–466).

Set S is {0, 0, 0, 2, 2, 4, 4, 4, 5, 5, 5, 5, 5, 7, 9, 9, 9, 9, 11, 11, 11}, the same set used in the discussion for correlation strength above (namely, all notes for ±4 at RU 462 of *White on White*). For PC 4, for example, $c_4(S)$ would be 3, as there are three instances of that PC in the set. S exerts the force shown in Fig. 2.19 on Fourier Balance 2 (which is the balance that will be used as an illustration for this discussion).

The magnitude of the arrow for PC p corresponds to $c_p(S)$, namely if there are four occurrences of PC 9, the length of the arrow is 4. (A length of 1 equals the length of the panhandle in the illustration, as demonstrated by the arrow for PC 7.) In this illustration, force lines have been placed slightly adjacent to (rather than directly on) the panhandles to avoid superimposing lines for PCs that share a pan (such as 5 and E).

To make the calculations for total resulting force, these individual forces are considered vectors on a Cartesian plane. In order to determine the resulting force f, a

---

[5] When using radians, $2\pi = 360°$, meaning that $30° = \pi/6$.

Fig. 2.19 Etude 15, RUs
458–466, PC mapping on
Fourier Balance 2 (the
lengths of the arrows are
proportional to the
cardinality of each PC
subset, where a cardinality
of 1=the length of one
panhandle)

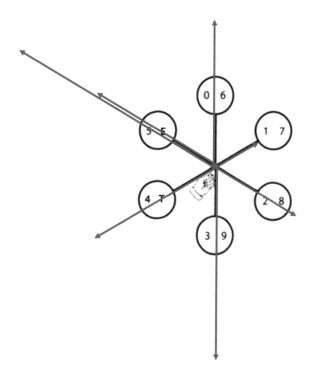

few trigonometric functions are used. First, the individual vectors are broken down into x and y components—Fig. 2.20 illustrates this for $c_2(S)$. To determine the x or y component of the vector for p, two pieces of information about it are needed: the magnitude and the direction. As mentioned, the magnitude is already given by $c_p(S)$, which in the case of $c_2(S)$ is 2.

The direction will be given by the angle a. Figure 2.21 illustrates a preliminary step for how the angle is calculated, using Fourier Balance 1. The plane along PC 0 is our starting point—angles are calculated clockwise from the y-axis, rather than counterclockwise from the positive x-axis (as in usual procedures on a Cartesian plane). As shown in Fig. 2.21, to move from 0 to 1 one makes a hop of one step (from one pan-handle to the next, clockwise), from 0 to 2 two hops, 0 to 3 three and so forth—the number of hops is equivalent to the PC. Each hop is 1/12th of the total number of degrees around the circle, namely $360°/12 = 30°$. Thus, the angle of two hops is $60°$, three $90°$, and so forth.

The relationship of the balance number and the manner of hopping becomes evident as soon as the balances are redrawn with the 12 pan-handles still intact, as shown in Fig. 2.22 for Fourier Balance 2. Here, to move from 0 to 1, one has to move two $30°$ hops, from 0 to 2 four, 0 to 3 six, and so forth. Hence, we have multiplied the PC p not only by $30°$ but also by two, in order to get the angle of the vector for p.

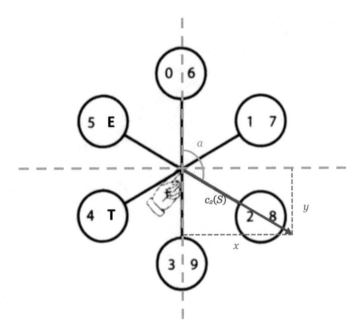

Fig. 2.20 Fourier Balance 2, mapping of PC 2 with a cardinality of 2 (includes Cartesian plane reference points)

Fig. 2.21 Fourier Balance 1, angle breakdown

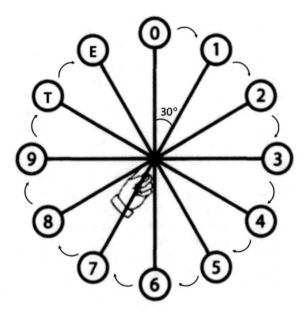

Fig. 2.22 Fourier Balance
2, redrawn around
12 panhandles

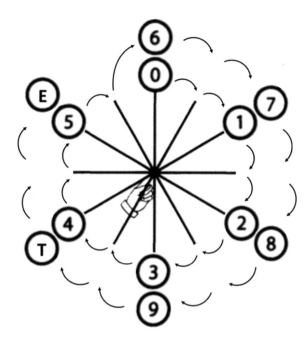

A pattern emerges. A quick look at Fourier Balance 3 (Fig. 1.3) will show that moving one PC requires *three* clockwise 30° hops—and the same pattern goes for the rest of the balances. Hence, *a* for vector *p* on balance *b* is 30°*p*b.[6] This may also be expressed in radians instead of degrees as $\frac{\pi}{6}pb$.

Knowing the vector magnitude and angle, *sin* and *cos* functions can be used for the x and y components (these are inverted as we are measuring angles from the y rather than x-axis). Namely, magnitude*sin(angle) gives the x component, and magnitude*cos(angle) the y:

$$x = c_p(S) \sin\left(30^\circ pb\right) \text{ or } x = c_p(S) \sin\left(\frac{\pi}{6}pb\right)$$

$$y = c_p(S) \cos\left(30^\circ pb\right) \text{ or } y = c_p(S) \cos\left(\frac{\pi}{6}pb\right)$$

In order to calculate the total force, the x and y components are each summed. This essentially creates one "big" triangle that is a summary of the individual vectors. The summing process for the PCs *p* ranging from 0 to 11 is expressed as follows (using radians):

---

[6]This is also explained, though in somewhat different terms, by Quinn's Multiplication Principle: "on Fourier Balance *n*, the PC *p* is located at (n*p) o'clock, where *n* and *p* are integers modulo 12 and should be multiplied accordingly" (Quinn 2007, 36).

$$\text{sum of } x \text{ components} = \sum_{p=0}^{11} c_p(S) \sin\left(\frac{\pi}{6}pb\right)$$

$$\text{sum of } y \text{ components} = \sum_{p=0}^{11} c_p(S) \cos\left(\frac{\pi}{6}pb\right)$$

The total force is the hypotenuse of this triangle (the sums of x and y components forming the two other sides at a right angle) for which we use the Pythagorean Theorem:

$$f = \sqrt{(\text{sum of } x \text{ components})^2 + (\text{sum of } y \text{ components})^2}$$

Hence, the total force f on balance b from set S is as follows:

$$f_b(S) = \sqrt{\left(\sum_{p=0}^{11} c_p(S) \sin\left(\frac{\pi}{6}pb\right)\right)^2 + \left(\sum_{p=0}^{11} c_p(S) \cos\left(\frac{\pi}{6}pb\right)\right)^2}$$

In the case of the current set S (± 4 span at RU 462) on balance 2, the result of this equation is 6.1 Lewins (as shown in Fig. 2.23b). While a number of conclusions can be drawn from a single result for correlation strength (as it already reveals some internal proportions of the set of notes), for a scalar force figure a certain amount of context is needed; it becomes highly valuable when compared to RUs before and after it or to the force from other spans at the same point—these are the variables displayed in the individual force graphs (Figs. 2.23, 2.24 and 2.25)—as well as a comparison with the force this moment exerts on different balances, which can be read either across the six force graphs for all spans or on the final force comparison graph (Fig. 2.26) with one representative span (in this case ± 6).

The dominant scalar force in the conclusion of *White of White* is predictably the diatonic, which soars well above the rest for the majority of the section (see Fig. 2.26). The diatonic force also shows a progressive weakening through the section brought about by the black-note introductions (Fig. 2.25a), with the main dips corresponding with those in the correlation strength (namely the chromatic episodes around RUs 510–511 and RUs 546–592).

However, unlike the correlation strength the diatonic force does not manage to fully recover from the first chromatic introductions (at RUs 510–511) until the next ones. This is due not only to the entry of the first black notes, but also to the nature of the ensuing white-note passage (Fig. 2.27), which causes the diatonic force to continue its precipitous plunge as well as a spike of the tritone force to a level above any other scalar force (Figs. 2.23b and 2.26). For the ± 6 span, the peak of the tritone force coincides with the bottom of the diatonic force dive at RU 527.

To clarify this pattern it is worth observing the mode breakdown for RUs 521–533 (namely the ± 6 span for RU 527). Where S_n is the set for the nth mode,

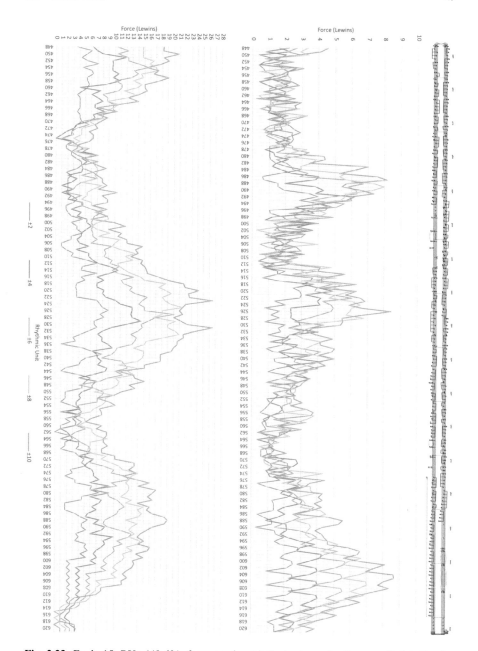

Fig. 2.23 Etude 15, RUs 448-621, force graphs. (**a**) (top): chromatic (force applied to Fourier Balance 1, measured in Lewins—span comparison). (**b**) (bottom): tritone (force applied to Fourier Balance 2, measured in Lewins—span comparison)

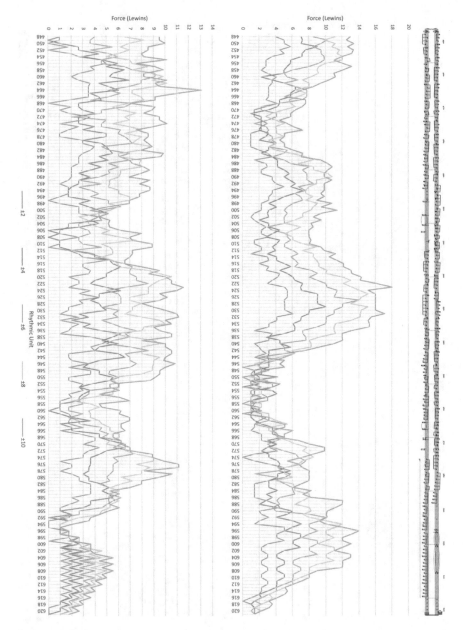

Fig. 2.24 Etude 15, RUs 448-621, force graphs. (**a**) (top): hexatonic (force applied to Fourier Balance 3, measured in Lewins—span comparison). (**b**) (bottom): octatonic (force applied to Fourier Balance 4, measured in Lewins—span comparison)

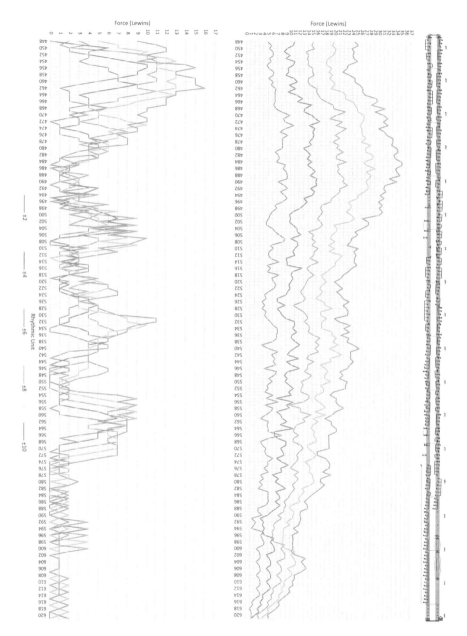

Fig. 2.25 Etude 15, RUs 448-621, force graphs. (**a**) (top): diatonic (force applied to Fourier Balance 5, measured in Lewins—span comparison). (**b**) (bottom): whole-tone (force applied to Fourier Balance 6, measured in Lewins—span comparison)

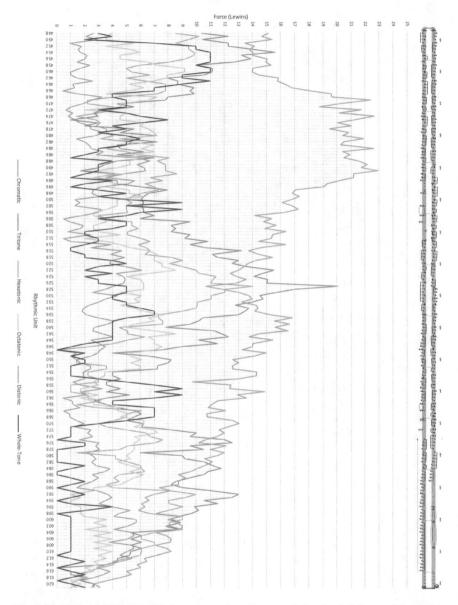

Fig. 2.26 Etude 15, RUs 448–621, force comparison (force applied to the six balances, using ±6 as a representative span)

$S_1 = \{5, 5, 5, 5, 5, 5, 5, 5\}$; $|S_1|=8$
$S_2 = \{11, 11, 11, 11, 11, 11, 11\}$; $|S_2|=7$
$S_3 = \{0, 0, 0, 0, 0, 0, 0\}$; $|S_3|=7$
$S_4 = \{4, 4, 4, 4, 4, 4, 4\}$; $|S_4|=7$

Fig. 2.27 Etude 15, RUs 506–536, score. © 2005 Schott Music, Mainz—Germany

$S_5 = \{7, 7, 7\}; |S_5|=3$
$S_6 = \{9\}; |S_6|=1$

As mentioned in the methodology, Fourier Balance 5 is only partly reflective of diatonic force: the diatonic scale registers a weaker force than the 024579 hexachord, given the presence of a tritone (namely, the fourth and seventh degrees in a major scale) which are on diametrically opposite panhandles and hence cause a balancing out. As shown by the mode breakdown, in this passage there are four PCs that are prominent: F is the most prominent at eight iterations, followed by C, E, and B tied at seven; F and B are the tritone in the white-note diatonic scale and thus cancel each other out. (The other two PCs in this passage are very infrequent—G at three occurrences and A appearing once.) The main PCs in this passage cause a weak diatonic force at 10.6 Lw—a level only reached again at RU 557, during the next chromatic passage.

On the other hand, these four PCs cause a considerable tipping on Fourier Balance 2 as they occupy adjacent panhandles within 180° on the balance—yielding 20 Lw, nearly double that of the diatonic force. The effect of these four PCs on balances 2 and 5 is illustrated in Fig. 2.28 (which, for the simplicity of the visual representation, places PCs in two categories of cardinality—the four frequent PCs are in bold colors, while the two rarer ones are in light shades). While PCs 0 and 4 are in the same positions on both balances, on balance 5 PCs 5 and 11 are at 180° and thus cancel out, while on balance 2 they are at 0°, hence applying force in exactly the same direction.

This example shows how even a combination of correlation strength and PC graphs can be limited in their use. While all the spans for 7-PC sets shot back up to full correlation in this passage, the data cannot reflect the nature of the PC set there is a correlation to. For this one may use the PC graphs (in particular the Co5 ordering), but even there, it is difficult to derive many conclusions beyond what the general PC set is. For example, the passage in question (RUs 521–533) seems to occupy the white-note set much in the same manner as for example RUs 485–497 (though compared to it seems a bit patchier)—there is not much to indicate that the latter (but

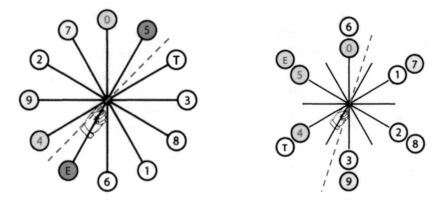

Fourier Balance 5 (diatonic force)   Fourier Balance 2 (tritone force)

Fig. 2.28 Etude 15, RUs 521–533, mapping on Fourier Balances 5 and 2 (tipping caused by PCs in the passage; green marks complementary PCs, red opposing; primary PCs in bold colors, secondary ones in light shades)

in the course of the work, earlier) passage has a diatonic force of more than twice the strength at 22.9 Lw.

These differences in graph representations are illustrated by Fig. 2.29: the white-note passages on either side of the chromatic moment at RUs 510–511 seem largely similar in the PC and correlation strength graphs, but the force graph shows a considerable change. It is thus the sensitivity of the force graphs to harmonic quality that make them an essential complement to the other two graph types. (That is not to say that they are enough alone: for example, a transposition of the entire set of notes analyzed would yield the same forces, but the PC graphs would paint a very different picture.)

For this reason that it is worth taking a look at the comparison graph for the entire *Vivacissimo* in Fig. 2.30: the force graph is the only type of graph that is able to distinguish between harmonically different types of white-note sets, while the PC and correlation strength graphs stay static, as expected.

Figure 2.30 is an important verification of the idea that the textural change at RU 448—moving from clusters in diatonic PC space to those in harmonic PC space—has a significant change on the harmonic quality of the macroharmony, which in terms of the overall PC set stays the same. The force graph shows that in the first part of the *Vivacissimo* (RUs 121–447) the diatonic force is generally the strongest but sometimes overshadowed by other forces depending on the PCs emphasized (such as a 579E cluster at RUs 240–252 for the strong whole-tone force). However, as soon as the primary verticalities become fifths at RU 448, the diatonic force begins to surge (as fifths occupy adjacent panhandles on Fourier Balance 5, causing a much stronger tipping than ICs more spread out—assuming identical cardinalities); the

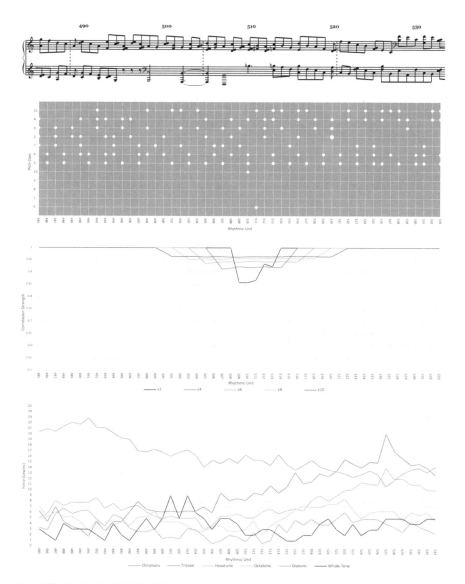

Fig. 2.29 Etude 15, RUs 485–533, comparison of three graphs types (below the notation row is the Co5 PC graph, followed by 7-PC correlation strength and finally the force comparison for ±6)

new passage is more consonant, which in the diatonic context renders the texture in a certain sense more "tonal."

This idea of tonality is only relative, as if the marker on an imaginary scale between vaguely-defined notions of atonality and tonality has moved a few steps towards the latter. However, this idea takes on greater importance later on in the passage with the appearance of a number of triads—the rarest type of simultaneity in

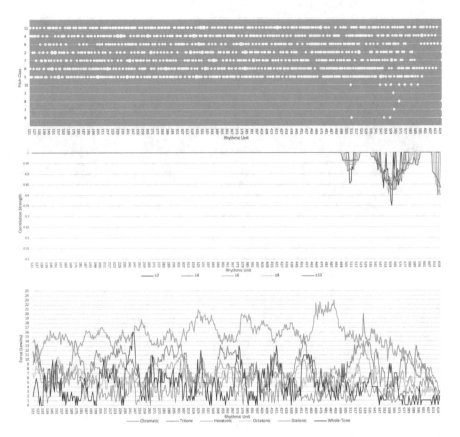

Fig. 2.30 Etude 15, RUs 121–621, comparison of three graph types (Co5 PC graph, 7-PC correlation strength and force comparison for ±6)

White on White; the presence of this chord type has significant implications for the larger framework of the etude.

For this it is necessary to briefly address the opening of the work. Triads do not appear at all in the first part of the *Vivacissimo*, and as mentioned above, the *Andante* is composed primarily of stacks of seconds and fourths—but not exclusively. There are two triads in the A-section of the *Andante* (which each get repeated given the ternary structure of the section). Only one of them, however, is a clean triad, i.e. without any other PCs in the same verticality (the other is part of a ninth chord G-D-F-A at RU 16)—the A-minor triad at RU 21 (Fig. 2.31).

A reason why this cannot be dismissed as a mere surface sonority is due to two features of this moment in pitch space. Firstly, the triad harmonizes the peak of the melody (the registral apex of the *Andante*), while secondly the distance between the bottom and top notes of the chord is at its greatest compared to the rest of the simultaneities in the section. It is also worth noting that unlike other white-note minor triads, this one in fact represents the key which the scale area belongs to—as if

Fig. 2.31 Etude 15, RUs 17–32, score. © 2005 Schott Music, Mainz—Germany

Fig. 2.32 Etude 15, RUs 49–64, score. © 2005 Schott Music, Mainz—Germany

in this moment of greatest registral stretching, a feature of the inner fabric of the texture is revealed.

This stretching is counterbalanced by the harmonization of this melody note in the B section at RU 61 (Fig. 2.32), where it is instead placed in a diatonic PC cluster (built upon the melody note in PC space, namely E-F-G-A).

Ligeti seemingly establishes two poles within the macroharmony, using these registral peaks as a signpost. One of them is a dissonant, non-tonal diatonicity— exemplified by the diatonic PC cluster—and the other is a more "harmonic" one, represented by the triad. It is precisely those two forces that have been in opposition throughout the piece, the first part of the *Vivacissimo* emblematic of the former, the conclusion of the latter—and the appearance of triads towards the end is a further reinforcement of this dichotomy (Fig. 2.33).

The most important triad in the conclusion is not one that appears as a simultaneity, but rather outlined in a broken chord. After the constantly moving and irregular figurations of the *Vivacissimo*, the final, insistent four-note figure in the LH (RUs 593–616) is remarkably anomalous. It is even more striking to consider that it actually highlights none other than an A-minor triad—the very same chord that marked the registral peak of the opening. Thus, the conclusion of the etude in fact refers back to the opening, creating a kind of harmonic recapitulation. The conflict between linear and harmonic diatonicity has been decided in favor of the latter—which, for good measure, is reinforced by the final chord, as if the Co5 skeleton upon which the conclusion is built is finally unveiled. (It is worth mentioning that the final chord is also the only one in the work to contain *two* triads within it, both of which frame the gap in pitch and harmonic PC space between the portions of the two hands: the higher frame is A-F#-C#, the lower frame the inverse D-A-F#.)

From a combination of the PC, correlation strength and scalar force graphs one can derive not only the composition of different macroharmonies, but the manner in

Fig. 2.33 Etude 15, RUs 538–621, marked score (triads and triad-containing chords highlighted in blue) © 2005 Schott Music, Mainz—Germany (unmarked)

which their content (in features considerably more detailed and nuanced than simply the scale area or PC sets they constitute) and juxtaposition create both musical structure and dramatic narrative. In pieces where macroharmony is arguably the primary feature (given the severe restriction of many other variables in dynamics, rhythm and so forth) that drives the listener's long-range perception of the works, such analysis proves especially useful. It will also serve to draw parallels across the etudes of Book 3, for which *White on White* serves in several ways as a kind of blueprint. One unusually specific and fascinating manifestation of this is in fact the pivotal nature of the A-minor triad, which will play a similarly important role in the rest of the set.

Chapter 3
Gradual Disintegrations in Etude 16, *Pour Irina*

3.1 General Remarks

Among the etudes of Book 3 it is the second one, *Pour Irina* ("For Irina"), that is perhaps the most reflective of the techniques Ligeti adapted from chaos theory in mathematics and often deployed in his music—namely, the introduction of a slight disturbances in a system that cause it to eventually spiral out of control. While this chaos principle is a primarily rhythmic one in his earlier work, in Etude 16 Ligeti applies it to scale area as well as tempo in a novel manner that makes the piece perhaps the greatest outlier among all of the 18 etudes. The central claim of this chapter is that *Pour Irina* is primarily a study in departure from the framework with which it opens; there is never a return home, but rather an arrival at a distant destination. This is effected by a number of processes through the etude, and each section approaches this question of disintegration in unique ways.

While one could argue that the end of *White on White* is also an example of a gradual scalar disintegration—i.e. the undermining of the strict white-note system—*Pour Irina* goes much further. As seen in the previous chapter, in the conclusion of Etude 15 there are only a few foreign notes that cause more than a gentle fuzzification of the dominant macroharmony. In Etude 16, not only is there a complete departure from the opening scale area but also a large-scale movement in harmonic space; Ligeti effects this by starting in a flat key area, moving through the white-note set and eventually landing in a sharp key.

As in the previous etude, *Pour Irina* begins with a lengthy passage that adheres completely to the opening—and essentially diatonic—macroharmony. While this process already seemed quite radical in *White on White*, Etude 16 takes it a step further, given the choice of scale area. There is something quite automatic or "natural" about the particular diatonic macroharmony of Etude 15, given that it is simply the white-note set; it is a scale immediately presented by the spatial arrangement of PCs on a keyboard, and the way it is used in the work does not give a strong implication of a particular major/minor scale or mode. The opening macroharmony

N. Namoradze, *Ligeti's Macroharmonies*, Computational Music Science,
https://doi.org/10.1007/978-3-030-85694-6_3

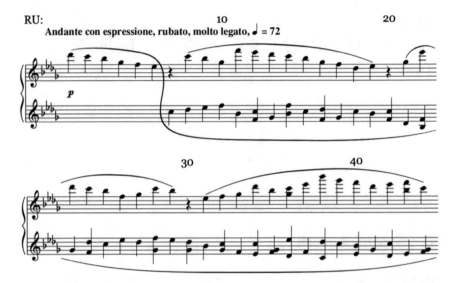

Fig. 3.1 Etude 16, RUs 1–44, score. © 2005 Schott Music, Mainz—Germany

in *Pour Irina* however is in a sense more specific in construction (being a particular selection of black and white keys, rather than just the white-note gamut) and much less anonymous in treatment, with a clear implication of a particular tonality. The notes of the first 164 RUs of the work adhere to only 6 PCs (C, Db, Eb, F, Gb and Bb) and while there are a number of contexts in which this set of PCs can be used, the shape of the motion in the beginning of the etude makes B-flat minor the clear, uncontested tonality. (The scale is simply a B-flat minor one with a missing Ab, hence the peculiar key signature in this opening as shown by Fig. 3.1.)

Despite the fact that, like in the opening of *White on White*, there is an avoidance of clean triads (though there are many triads embedded in other, primarily seventh, chords), this opening *Andante con espressione* is arguably the closest of any passage of Ligeti's music (save, perhaps, for a few works written in his very first years of composition) to common-practice, functional harmony, each simultaneity colored by our perception of "harmony/non-harmony" notes and chordal functions. It is perhaps in order to highlight the ensuing process of scalar chaos that the territory of order is so stable—not only in macroharmonic purity but also in tonal anchoring. In this harmonic context, the largely-stepwise moving, highly melodic upper line, with phrases articulated by rests, creates a style of expressivity unique in Ligeti's oeuvre.

The title *Pour Irina* (named after its dedicatee, the pianist Irina Kataeva, who premiered the work) in fact brings to mind another work of music that this opening *Andante* may allude to—the solo piano piece *Für Alina* ("For Alina") by Arvo Pärt (b. 1935). As with *White on White*, we have no written record of whether the title for Etude 16 does in fact reference the Pärt work or was at all influenced by it, but there are certainly a number of musical parallels: the instrumentation, sparse and medita-tive texture and minor key signature (which are only a half-step apart, as *Für Alina* is

Fig. 3.2 Arvo Pärt, *Für Alina* (1976), bars 1–7, score. © Copyright 1990 by Universal Edition A. G., Wien

in B minor). In fact, there are further similarities to *Für Alina* in the very beginning of the etude (namely the first two dozen RUs or so), not only in the registration of the moving voices but, more importantly, in the fact that the "bass notes" (the occasional third voice at RUs 12, 14 and 16, for example) are all notes of the B-flat minor triad while the others use largely stepwise motion. This is an almost direct reference to Pärtian tintinnabulli, the very definition of which is diatonic stepwise motion in one voice with tonic-chord arpeggiation in the other (Hillier 1997, 99), as can be observed in Fig. 3.2. Ligeti's use of such highly restricted parameters at the outset help to magnify the scale of the ensuing journey.

This departure from the framework with which *Pour Irina* opens is highlighted by tempo changes over the course of the work, a stepwise acceleration that reflects the steadily progressing disorder and agitation. The *Andante* is followed by an *Allegro con moto* where the RUs are in running eighth-notes; then comes an *Allegro vivace* (marked as 1.5 times faster, so that three eighth-notes equal two eighth-notes from the previous texture) and finally a *Molto vivace* (in 16th notes, four of which are equal to three eighth-notes in the *Allegro vivace* and two in the *Allegro con moto*; thus, this final section essentially has the same tempo as the *Allegro con moto*, just with RUs twice as fast). Of Ligeti's previous etudes, only one had more than two different tempi over the course of the work (namely Etude 3, *Touches bloquées*— "Blocked Keys"—where the central *feroce* section has two slightly varying degrees of forward motion); *Pour Irina* has four.

Each of these four sections engage in the broader overall movement towards disorder in different ways, with features including scalar correlation, modulation and interval relationships. The changes in these variables are accentuated by their divergent treatment in the two hands. In fact, the two hands have a greater degree of independence in *Pour Irina* than in any other etude of Book 3, and the extent to which they are (or are not) in sync is another feature that will affect our perception of degrees of stability (much in the manner Ligeti would have previously used this principle for rhythmic divergence, such as in the aptly named Etude 1, *Désordre*— "Disorder").

The graphical analyses prove to be powerful tools in reflecting these various destabilizing processes, both on a local and broader level. Large-scale changes are evident, for example, in the broad decline in correlation strength (Figs. 3.7, and 3.8) and diatonic force (Fig. 3.11a) over the course of the etude. In fact, the Co5 PC graph (Fig. 3.4b) shows how *Pour Irina* is in a sense an inversion of *White on White*: while Etude 15 began at the top of the (for this paper) standard linear arrangement of the Co5 and then encroaches upon the lower portion (Fig. 2.8b), Etude 16 begins at the bottom and then proceeds to expand upwards. The commentary will also address the smaller-scale features the graphs can reflect, such as intervallic relationships or how the sections differ in the quality of their macroharmonic motion. These reveal another manner in which Etude 16 is a counterpart of Etude 15: while the majority of *White on White* was largely focused on "dissonant diatonicity," in the central sections of *Pour Irina* Ligeti inverts this to create "consonant chromaticism," where it is a consonant interval that is strictly controlled while the texture is allowed to ramble through the chromatic gamut.

3.2   Graphical Analysis (Figs. 3.3, 3.4, 3.5, 3.6, 3.7, 3.8, 3.9, 3.10, 3.11 and 3.12)

Fig. 3.3 Etude 16, RUs
1–598, piano roll (notes in
P-space, C3 = middle C)

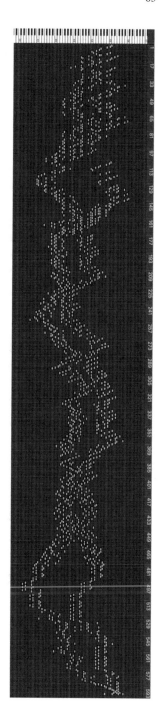

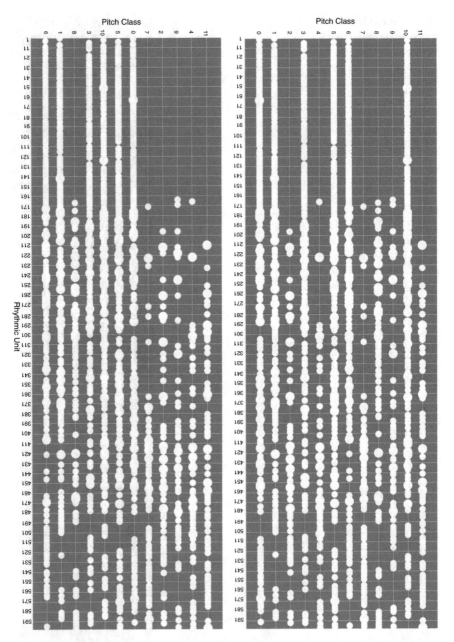

Fig. 3.4 Etude 16, RUs 1-598, PC graphs. (**a**) (top): chromatic (notes placed in PC space ordered chromatically). (**b**) (bottom): Co5 (notes placed in PC space ordered by the Co5)

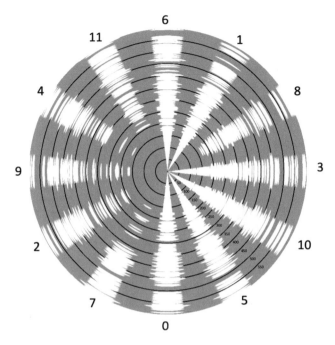

Fig. 3.5 Etude 16, RUs 1–598, conical Co5 PC graph (notes placed in PC space ordered by the Co5 using polar coordinates: the y-axis is rendered as a circle, and the black rings mark the passage along the x axis)

3.3 Commentary

The opening six notes of Etude 16 are a descending scale that presents all six PCs of the scale area. This PC set—01356T—could be considered either a diatonic scale with a missing PC (in this case 8) or an extension of a minor hemitonic pentatonic scale such as the Japanese Hirajōshi scale. The *Andante con espressione* is 175 RUs in length, and there is complete adherence to this six-PC scale through the first 164 RUs.[1] This is reflected by the total correlation in the six-PC correlation strength through this opening as shown in Fig. 3.14.

As can be expected, the example shows that the generally dominant force throughout this passage is the diatonic. What is somewhat more surprising is the fact that the tritone force is a close second, and at a few moments even the strongest force; this is not a pattern one would expect in a (broadly-speaking) diatonic scale area. Interestingly, the hierarchy of forces is relatively consistent throughout: the diatonic is the strongest overall, followed by the tritone, then the octatonic, hexatonic, chromatic and finally whole-tone. This indicates that there is a generally

---

[1] The term "hexatonic" is avoided here, so as to not cause confusion with the hexatonic scale type associated with Fourier Balance 3.

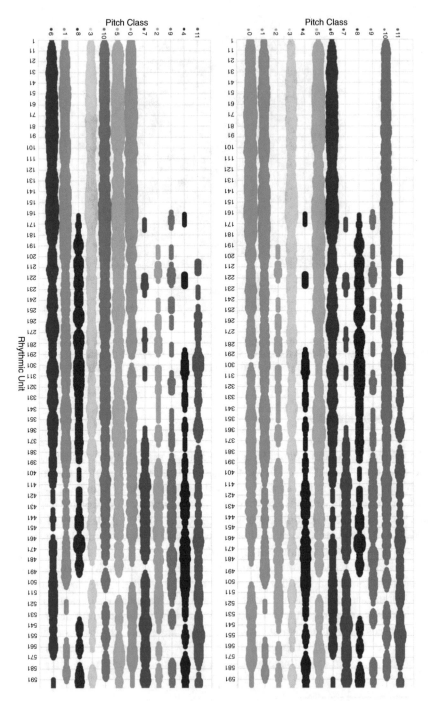

Fig. 3.6 Etude 16, RUs 1-598, PC graphs. (**a**) (top): chromatic, ±6 span (notes placed in PC space ordered chromatically, with a span of ±6 applied to each RU). (**b**) (bottom): Co5, ±6 span (notes placed in PC space ordered by the Co5, with a span of ±6 applied to each RU)

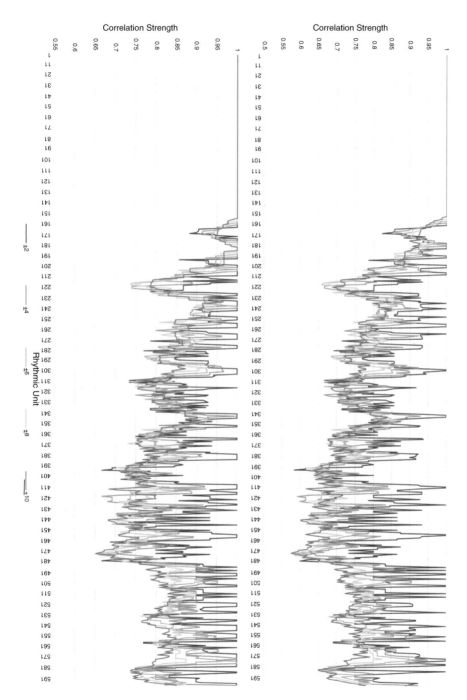

Fig. 3.7 Etude 16, RUs 1-598, correlation strength graphs. (**a**) (top): 6-PC (number of notes that belong to the 6 most frequently appearing PCs/total number of notes). (**b**) (bottom): 7-PC (number of notes that belong to the 7 most frequently appearing PCs/total number of notes)

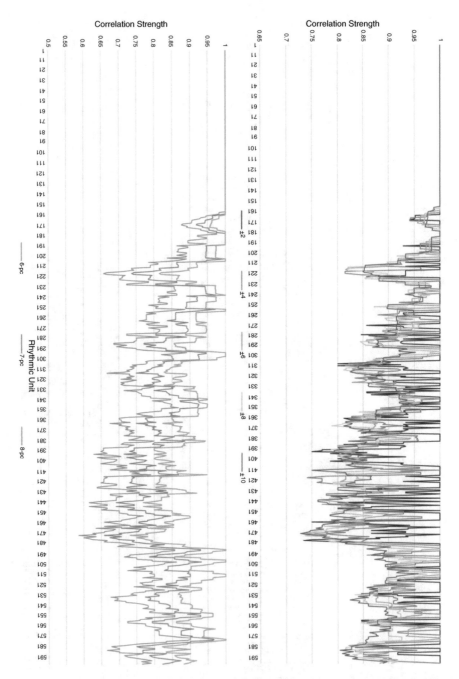

Fig. 3.8 Etude 16, RUs 1-598, correlation strength graphs. (**a**) (top): 8-PC (number of notes that belong to the 8 most frequently appearing PCs/total number of notes). (**b**) (bottom): comparison (comparison across 6-, 7- and 8-PC sets using ±6)

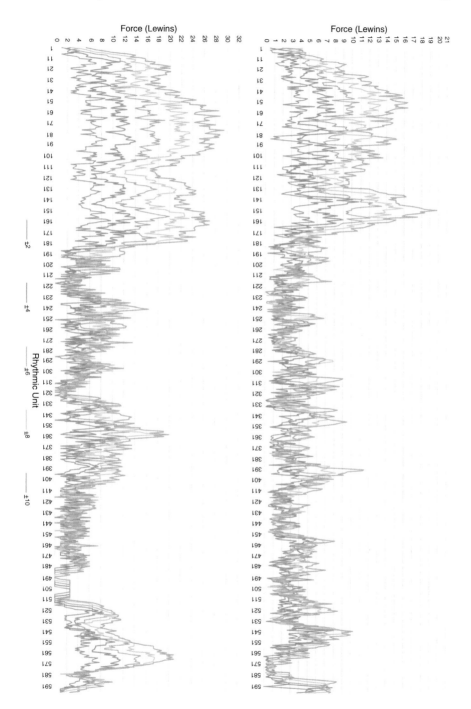

Fig. 3.9 Etude 16, RUs 1-598, force graphs. (**a**) (top): chromatic (force applied to Fourier Balance 1, measured in Lewins—span comparison). (**b**) (bottom): tritone (force applied to Fourier Balance 2, measured in Lewins—span comparison)

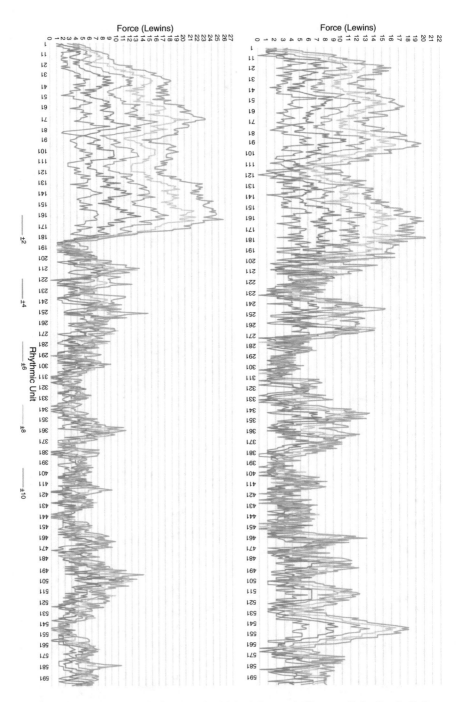

Fig. 3.10 Etude 16, RUs 1-598, force graphs. (**a**) (top): hexatonic (force applied to Fourier Balance 3, measured in Lewins—span comparison). (**b**) (bottom): octatonic (force applied to Fourier Balance 4, measured in Lewins—span comparison)

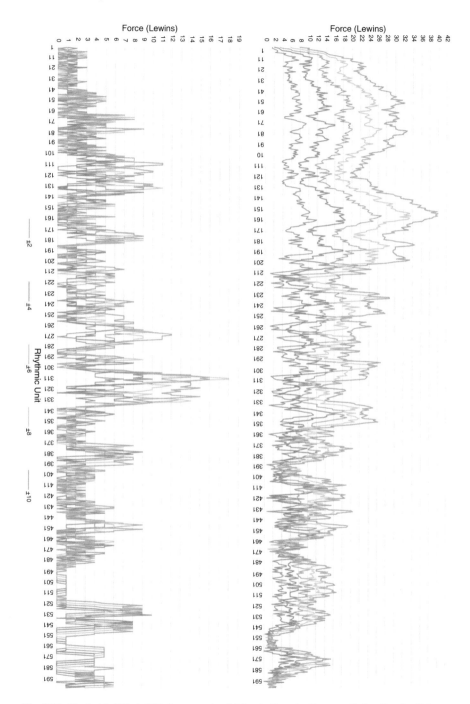

Fig. 3.11 Etude 16, RUs 1-598, force graphs. (**a**) (top): diatonic (force applied to Fourier Balance 5, measured in Lewins—span comparison). (**b**) (bottom): whole-tone (force applied to Fourier Balance 6, measured in Lewins—span comparison)

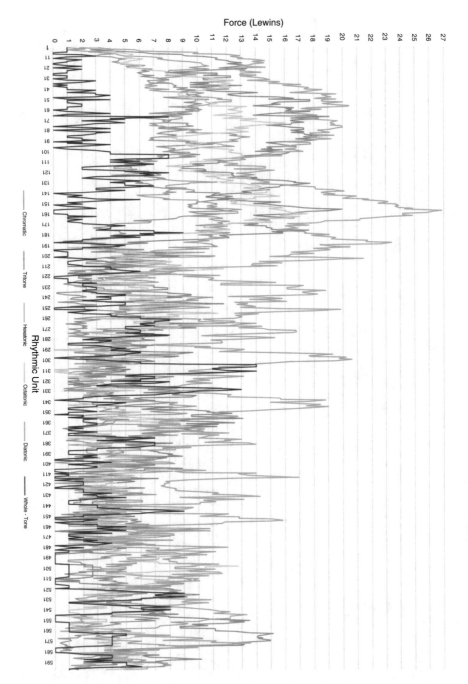

Fig. 3.12 Etude 16, RUs 1–598, force comparison (force applied to the six balances, using ±6 as a representative span)

Table 3.1 Force (in Lewins) on each Fourier Balance from different six-PC diatonic subsets

Missing PC from (013568T)	Fourier Balance					
	1	2	3	4	5	6
None (full set)	0.27	1	1	1	3.73	1
0 or 6	1.04	0	1.41	0	3.86	0
1 or 5	0.9	1	0	1.73	3.35	2
3	0.73	2	2	0	2.73	2
8 or 10	1.24	1.73	1.41	1.73	2.91	0

regular proportion of PC appearances through the *Andante* (given that the graph shows relatively little skewing or variance in the hierarchy, which would result from a greater concentration of a few PCs in certain passages), and that this proportion could perhaps simply be a relatively even appearance of all PCs of the scale—namely, that this pattern of scalar force strength could be something inherent in this six-PC scale.

It is thus worth simply plotting the scale on each Fourier Balance and recording the resulting force reading, as shown in Table 3.1. For reference, the full diatonic set is included, as is its every possible six-note subset, each labelled by the PC missing from 013568T (which is incidentally not only the full B-flat minor scale, but also the prime form for the diatonic scale). The subsets are grouped according to set class (given that, for example, 03568T and 01368T—the "missing 1" and "missing 5" sets—are equivalent in prime form, which is namely 023579).

The set being investigated in the opening of *Pour Irina* is the "missing 8" set, and Table 3.1 goes some way towards explaining the force readings in the *Andante*: the tritone force ties for second place alongside the octatonic behind the diatonic force, and these three forces are considerably closer to each other in the "missing 8" than in the full diatonic set. The hexatonic comes in fourth, followed by the chromatic and whole-tone—where there is in fact no force, as the set contains three PCs each from both whole-tone scales. In the full diatonic scale, on the other hand, the whole-tone force is considerably stronger than the chromatic, and the hexatonic is just as strong as the tritone and octatonic.

While Table 3.1 sheds light on the force readings in Fig. 3.14, it does not fully explain the strength of the tritone force (which in the table ties with the octatonic, rather than being the clear runner-up to the diatonic). It is hence clear that while the relative proportion of PCs is quite constant through the *Andante*, it is not one of simply equal appearance of each PC—there is a slight disbalance that yields this particular force-hierarchy.

Thus, it is worth simply summing the number of appearances for each PC through the whole *Andante* (including the foreign ones that appear in the 11 RUs of the section, to be addressed further on) to reveal the proportion of PC presence in the texture (Table 3.2) and then plotting this set of 509 notes on each Fourier Balance (Table 3.3).

Table 3.2 reveals that of the diatonic subset, PCs 0, 5 and 6 are around a third more prevalent than 1, 3 and 10—and it is this proportion that yields the strong

Table 3.2 Etude 16, RUs 1–175, number of appearances of each PC

PC	0	1	2	3	4	5	6	7	8	9	10	11
Frequency	100	75	0	71	2	94	91	1	1	2	72	0

Table 3.3 Etude 16, RUs 1–175, force for set of all notes plotted on each Fourier Balance

Balance	1	2	3	4	5	6
Force (Lw)	95.18	183.31	115.76	149.3	222.94	23

tritone force, as shown in Table 3.3. This relative ordering of forces remains in place even at the general peak at RUs 157–160, where the sheer volume of notes (due to the dominance of four- and five-note chords, in place of a previous focus on two- and three-note ones) cause a swell in most of the forces (except for the whole-tone force—balance 6—where more notes simply continue to cancel each other out), as shown in Fig. 3.14. It is interesting to observe the extent to which the increased volume of notes serves to essentially magnify the disparities between the forces, benefiting those that respond more sympathetically to this scale (namely those for balances 4–6) while not changing much for balances 2 and 3, all in a relatively proportional manner.

The broad hierarchy of forces does not account for the entire *Andante*, and some moments where this trend is broken are particularly revealing. One such example is the diatonic nadir at RU 111, an instance where the tritone force is in fact (narrowly) the strongest, a result of the particular dominance of PCs 0, 1 and 6. Here, PC 1 replaces 5 in prevalence, given that it is the supporting fifth above Gb (PC 6) at what is in fact a harmonic arrival at the submediant that precedes a recapitulation of the opening melodic line (Fig. 3.14)—and the significance of this moment is highlighted by the leap in the RH of over two octaves, the only such registral jump in the whole etude (as can be observed in the piano rolls, Figs. 3.3 and 3.21).

However, this reharmonization of the opening material is not the first such occurrence in the etude. There is another far closer to the outset of the work, at RU 46, where we are still very much in the tonic area; the slightly modified reiteration of the opening shape is underpinned by a grounding Bb bass note at RU 45 that provides an impulse for it (as shown above in Fig. 3.13). What ensues is a lengthy registral descent that eventually reaches the landing at the low Gb at RU 108 (Fig. 3.14)—a halfway-point to the eventual return to and final arrival at the tonic at RU 175 (Fig. 3.15).

It is in this final approach to the tonic that the texture strays from the rigid adherence to the six-PC diatonic subset. Five foreign notes appear, belonging to four PCs: E, G, Ab and A (appearing twice). The last three of these fit within a broader conception of a Bb minor scale: G and A in the ascending and Ab in the descending melodic (or natural) scales. Curiously enough, the arrival to these notes does indeed fit with these parameters: the G (RU 172) and two As (RUs 165 and 170) are approached from below, the Ab (RU 168) from above. The only other foreign

Fig. 3.13 Etude 16, RUs 1–67, score. © 2005 Schott Music, Mainz—Germany

note is the E, the inclusion of which heralds its later importance—it is a PC which is in fact the eventual destination of the etude.

Interestingly, the patterns that this final PC-set (i.e. including the foreign notes) form on the chromatic and Co5 PC rolls are closely related to one-another. From a linear chromatic standpoint, the set includes adjacent PCs 0 and 1, skips 2 and then fills the gamut from PC 3 through PC 10, while on the Co5 ordering, it occupies adjacent steps 4 and 9, skips 2 and then fills the range from 7 through 6—in both cases, PC 2 acts as a pivot between a stretch of two adjacent PCs on one side, and seven on the other (Fig. 3.16). From a scalar force perspective, this broadening of the pattern of the initial six-note set has the largest effect on the chromatic, weakening it considerably; while forces such as the diatonic or tritone also show a downward curve after the note-density induced peak (Fig. 3.14), they return to pre-swell levels, while the chromatic continues to plunge, given the decrease in chromatic packing that the new PCs effect. It is also in these final RUs of the *Andante* that the correlation strength finally weakens (Fig. 3.16).

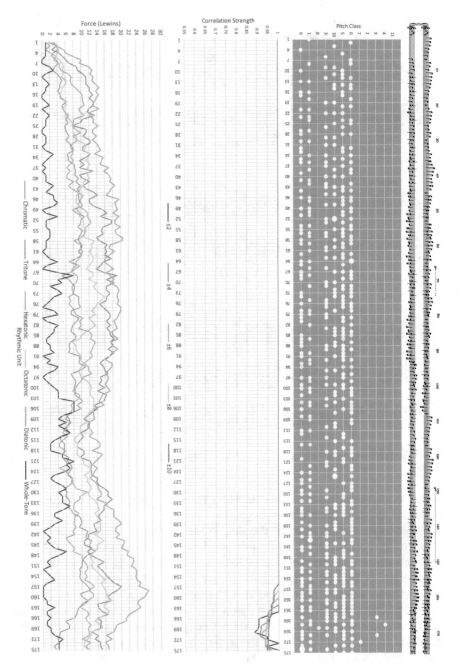

Fig. 3.14 Etude 16, RUs 1–175, comparison of three graph types (Co5 PC graph, 6-PC correlation strength and force comparison for ±6)

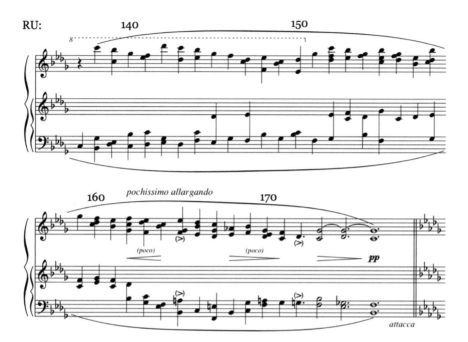

Fig. 3.15 Etude 16, RUs, 138–175, score. © 2005 Schott Music, Mainz—Germany

The primary purpose of these new PCs is to heighten the expressive impact of this final arrival at Bb in the *Andante*. This is not only a mere harmonic widening or departure from the strict adherence of the six-note set: the particular choice of foreign notes in each simultaneity serves to add greater emphasis and dissonance. The greater thickness of the chords (i.e. the increased note density) also highlights the significance of this moment, as do two performance markings: the swell in dynamics, which come to a peak at RU 165 (where in fact the first foreign note, an A natural, appears), and the broadening tempo, first in the *pochissimo allargando* and then the written-out ritardando (Fig. 3.15).

The final dotted half note *pianissimo* chord seems to encapsulate the essence of the entire section in a number of ways, making it a fitting landing spot for this section of remarkable poignancy. It contains both the tonic note as well as PCs 0, 5 and 6 which have been dominant throughout the *Andante*. However, not all of the notes are equally present. Given the fact that PC 6 is held over from the previous chords, the Bb, F and C are in the foreground for the listener, a relatively stable chord which creates a certain sense of arrival. On the other hand, the fainter, tritone-inducing Gb in the background is a distant sign that there is yet something unresolved, that a continuation awaits—much in the manner that most simultaneities in the *Andante* seem to be comprised of notes of the underlying harmonic function as well as suspensions/anticipations that thus avoid unambiguous triadic sonorities, creating a sense of perpetual motion and searching for a resolution. Despite the fact that at RU

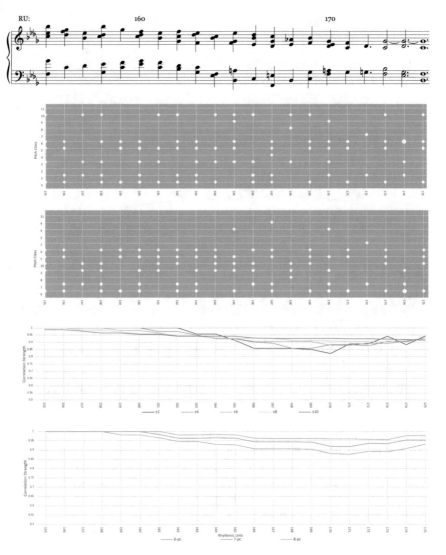

Fig. 3.16 Etude 16, RUs 155–175, comparison of four graph types (chromatic and Co5 PC graphs, 6-PC correlation strength and correlation strength comparison for ±6)

175 the texture has in a sense returned to the tonic and ground to a halt from a temporal perspective, a sense of anticipation still remains.

The continuation comes in the form of the ensuing *Allegro con moto*. From certain perspectives, it seems as if it largely picks up from where the *Andante* left off, given the similar (quiet) dynamic marking, register and key—we are still in Bb minor (with the scale firmly anchored on a recurring Bb as the bass note, lest there be any doubt), now filled by the Ab so we have the full diatonic set. However, the

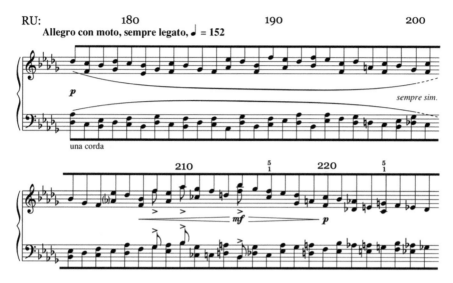

Fig. 3.17 Etude 16, RUs 176–227, score. © 2005 Schott Music, Mainz—Germany

texture at the outset of this new section is very different, mostly in three voices with all simultaneities forming IC 5; the focus on this intervallic composition lasts for 117 RUs, despite a few mostly accented outliers (such as RUs 208 and 210, shown in Fig. 3.17). Only at RU 292 does the general texture begin to move away from this strict framework.

The combination of full diatonicity and perfect-fifth predominance as the primary simultaneity creates an expected surge in diatonic force (Fig. 3.18) to a level that will never be reached at any point further on in the etude; the rest will be a general waning of diatonicity as well as correlation strength. In fact, the sections from the *Allegro con moto* onwards display a largely weaker showing than in the *Andante* for all scalar forces (save for the whole-tone, which fared particularly badly in the opening six-PC set) given the greater rate of macroharmonic change and chromaticism that ensues (see Figs. 3.7 through 3.12 in the graphical analysis).

This process of change begins at RU 197 with a D natural below an A natural (Fig. 3.17)—the very same first foreign PC as in the *Andante* (at RU 165). However, the integrity of the intervallic relationship in the simultaneity (namely that of IC 5) has remained intact—as it does in the ensuing foreign note intrusions at RUs 211–213, which almost complete the chromatic gamut. (The leftover PC, G natural, eventually appears at RU 224). Despite the quickly progressing chromaticism, the texture at the outset of the *Allegro con moto* generally remains consonant—as mentioned in the introduction to this chapter, a process that is very much the opposite of the dissonant diatonicity that defined the majority of *White on White*.

Another way in which the texture in this section—as well as in the *Andante*—serves as a counterpart to the processes of the preceding etude is the fact that the change in macroharmonic correlation is not organized by the Co5. As can be observed in Fig. 3.18, the PC graph shows no discernible Co5-related pattern in

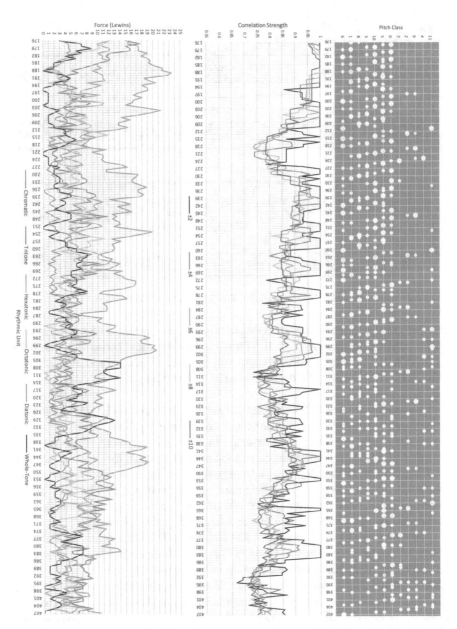

Fig. 3.18 Etude 16, RUs 176–408, comparison of three graph types (Co5 PC graph, 7-PC correlation strength and force comparison for ±6)

the introduction of foreign notes. In the *Allegro con moto*, the full chromatic set is reached early on (49 RUs in out of a total of 232); the rest of the section will merely show an increasing frequency of these foreign notes, as evidenced by the general decline in correlation strength (Fig. 3.18).

The primary manner in which the intervallic composition of the texture in the *Allegro con moto* changes is through the accented simultaneities, which tend to break the mold in ways that are then slowly adopted by the general passagework. For example, the accents at RUs 208 and 210 are seventh chords (ICs 1 and 2), intervals that later appear in non-accented simultaneities at RUs 257 and 269. Later, at RUs 266, 269 and 271, accented chords are the first verticalities with three PCs (each of them a seventh stacked on a perfect fifth, which when rearranged is in fact two stacked fifths)—and these chord types then appear in the non-accented texture from RU 307 onwards. At RUs 313 and 315, accented tetrachords introduce triads—a new kind of sonority, though, as in the *Andante*, they are not unobstructed and but rather embedded in chords with other PCs.

Another similarity with the *Andante* is the recurrence of the opening melodic shape (though, in the case of the *Allegro con moto*, it is somewhat less melodic) with underlying reharmonizations. The first such moment is when the upper line of RUs 179–213 appears an octave lower at RUs 230–262; considerably more important is the reappearance of a whole 53-RU stretch of the opening (RUs 176–228) at RUs 327–379 an octave higher, now *fortissimo* and with a far more dissonant harmonization (Fig. 3.19). Notably, the dissonance here is a result of changes in the LH only—the RH is identical to the opening (Fig. 3.17) in the upper line as well as in the configuration of all of the dyads (save for one missing F at RU 328).

Most of the simultaneities within the LH are still IC 5, although they no longer always align perfectly with those in the RH; this trend has become more pronounced through the *Allegro con moto* and comes to a head in the latter part of the section. Thus, this process of weakening macroharmonic correlation happens in tandem with (and is partly due to) a divergence *between* the two hands, an independence reflected by the perpetually shifting registral relationship of hands throughout the etude as shown by Fig. 3.20 (while the hands in almost all passages of the other etudes of Book 3 move across the keyboard largely in parallel). It is thus worth separating the data for the two hands in this section, in order to determine the extent to which each hand affects the global macroharmonic shift, as shown in Figs. 3.21 and 3.22.

While it is clear in the graphs for both hands that the PCs of the B-flat minor set are still the primary, dominant ones through most of the passage, they are far more present in the RH than the LH—a difference reflected in the correlation strength, which for the LH becomes far weaker in the passage than for the RH. This trend is also evident in the considerably stronger diatonic force in the RH than in the LH. In fact, the diatonic is consistently the strongest force type in the RH, while in the LH every other scalar force has a moment of dominance over the rest.

The tension that accompanies the progressing instability through the *Allegro con moto* is heightened by the swelling dynamics, which reach *fortississimo* at RU 397, near the end of the section. The focus on IC 5 has all but disappeared upon arrival to this climax in volume (for example, certain stretches such as RUs 387–396

Fig. 3.19 Etude 16, RUs 306–363, score. © 2005 Schott Music, Mainz—Germany

Fig. 3.20 Etude 16, RUs 1–598, piano roll (repeat of Fig. 3.3)

are comprised of parallel seconds), but at the very end there is a sudden cleanup, as the texture quickly reverts back to the original configuration (Fig. 3.23)—a realignment that prepares for the ensuing *Allegro Vivace*.

As discussed above, several features serve to smoothen the transition from the *Andante* to the *Allegro con moto*, such as register, key and dynamics—as if to

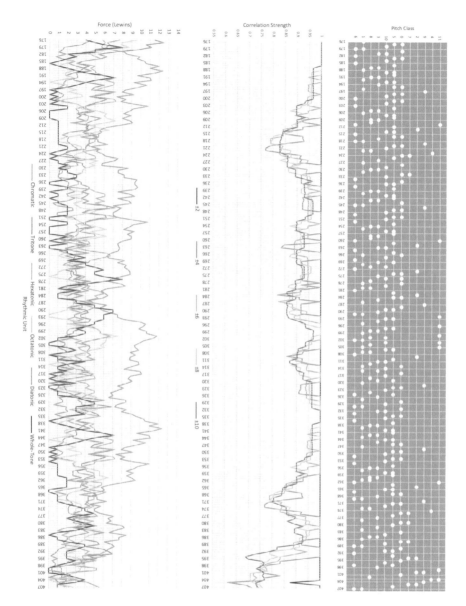

Fig. 3.21 Etude 16, RUs 176–407, RH, comparison of three graph types (Co5 PC graph, 7-PC correlation strength and force comparison for ±6)

mitigate the jump in tempo (made even more stark by the deceleration at the end of the first section). A similar process occurs between the *Allegro con moto* and the ensuing *Allegro Vivace* but with a different configuration of variables. While the register remains the same as it had been in the earlier transition (the *Allegro vivace*

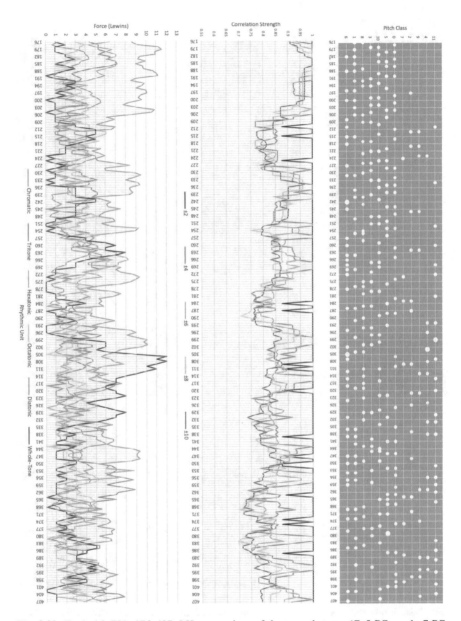

Fig. 3.22 Etude 16, RUs 176–407, LH, comparison of three graph types (Co5 PC graph, 7-PC correlation strength and force comparison for ±6)

begins with the same chord type with which the *Allegro con moto* ended, moved up a semitone), the dynamics are now instead a vehicle for contrast, given that there is a sudden shift from *fortississimo* to *pp*.

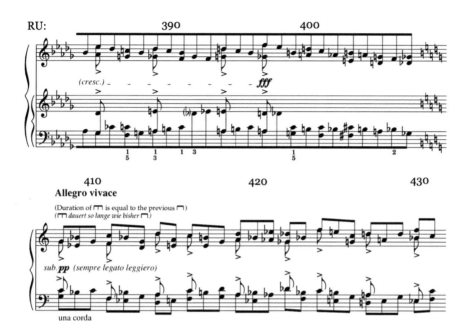

Fig. 3.23 Etude 16, RUs 384–431, score. © 2005 Schott Music, Mainz—Germany

A feature that serves as important glue at this transition is the intervallic makeup of the simultaneities, which also serves as the main contrast between the first two sections. The texture of the *Allegro vivace* starts off with largely the same features as those in the *Allegro con moto*: verticalities of IC 5 in three voices. However, the generally smooth, stepwise motion of the outer voices from the previous sections of the work has now been replaced by a far more angular one, underlined by accents for every two or three RUs that create a jaunty, Bulgarian rhythm-like pattern.

The most significant trend that has continued through all three sections is the steadily increasing rate of macroharmonic destabilization—and its harmonic direction. What begins as a few foreign notes in *Andante* (a gentle fuzzification of the otherwise undisturbed six-PC scale area) turns into a far more significant weakening of correlation in the *Allegro con moto*. There is also a particular direction for this process, which is a motion *up* the Co5. While the various Co5 PC graphs (Figs. 3.4b, 3.18, 3.22 and 3.23) all display this upward movement, Ligeti is in fact very specific about notating virtually all black notes as flats (there are only two exceptions, both in the LH at RU 339—see Fig. 3.19) and foreign notes as their natural cancellations. Thus, we are moving from flat territory upwards. This pattern continues in the *Allegro vivace*: the increasingly disorienting chromaticism that led into the section now morphs into rapid switches from one key area to another (another reflection of the new textural angularity), coupled with an initial oscillation between the white-note set and flatter key areas that eventually shifts upwards into sharp keys. Given the combination of macroharmonic instability and intervallic regularity, the *Allegro vivace* continues the consonant chromaticism that had characterized the texture in the *Allegro con moto*; as with the earlier section, the intervallic regularity eventually disintegrates as the end of the *Allegro vivace*.

At the outset of this new section, groups of four to six RUs—roughly coinciding with pairs of groups of RUs beamed together—adhere to diatonic, seven note scales, in the manner shown in Fig. 3.24. Their position on the Co5 can be charted with integer labels, relative to 0 for the white-note set; for example, the set 023579T (the PCs of the Bb-major scale) would be −2, and 024679E (the G major) would be +1. These labels assign the fewest number of switches that the PCs could imply, grouping the largest possible stretches of RUs under a single diatonic-scale umbrella. While some stretches of RUs could belong to more than one scale (namely if certain decisive scale degrees do not appear) the scales chosen are those as near as possible to the previous one in terms of Co5 steps; in places where the notes of certain RUs could belong to either the previous or the next scale area, the divisions are aligned with the rhythmic/beaming pattern wherever possible.

As with the increasingly frequent appearance of foreign notes in the previous parts of the etude, the shifts in scale area in the *Allegro vivace* also occur more often in the latter portion of the section. Figure 3.25 includes a graph that charts the progress of the diatonic scale areas on the Co5, in the manner used in Fig. 3.24.

As can be observed in Fig. 3.25, it is only towards end of the *Allegro vivace* that sharp key areas are reached: while the PC graph is unable to distinguish between

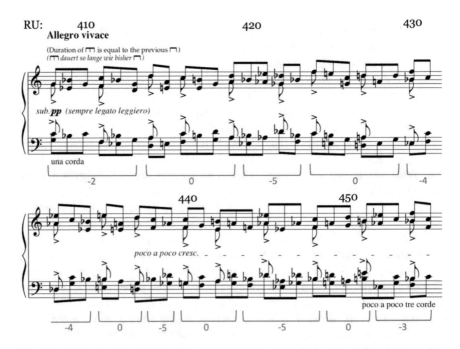

Fig. 3.24 Etude 16, RUs 408–428, annotated score (integers for Co5 scalar motion relative to 0 for the white-note set). © 2005 Schott Music, Mainz—Germany (unannotated)

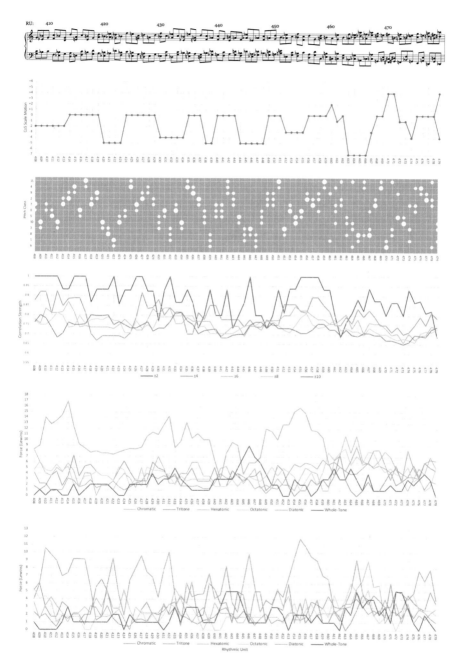

Fig. 3.25 Etude 16, RUs 408–479, comparison of five graph types (scalar motion, Co5 PC graph, 7-PC correlation strength, force comparison for ±6 and force comparison for ±2)

different spellings of the same black notes (which occupy the lower portion of this arrangement along the y-axis), the graph for the scale motion aligned above it can reflect Ligeti's particular spellings, revealing the final general upward movement. The roll is nevertheless very useful in showing the temporal organization of the PCs: it demonstrates the presence of Co5 motion even within and across each scale stretch. For example, the scale motion graph seems to imply that there is a sudden jump from 0 to −5 at RU 437–438, while the roll below it shows that there is in fact a much smoother motion down the Co5 starting at 435 through 439.

Smaller spans are particularly useful in the correlation strength and force graphs, given the rapid rate of switching between scale areas. For example, the ±2 data in the seven-PC correlation strength graph frequently rises to full correlation at the beginning of the *Allegro vivace* but then deteriorates as the rate of macroharmonic change increases over the course of the section (given that the scale stretches become smaller and are sometimes no longer than a single RU). Interestingly, there are four RUs of complete correlation at 455–458, which does not seem to agree with the scale area jump from −3 to 0 at RUs 455–456 shown in the scale motion graph. However, the Co5 PC roll shows a generally smooth upward motion in the area, meaning that no stretch of five consecutive RUs between RUs 453–460 contains more than seven PCs—another example of harmonic motion that smoothens or blurs the edges between these alternating diatonic PC sets.

In the usual ±6-span configuration, the force comparison graph shows a clearly dominant diatonic force in the opening portion of the section that then plunges when the rate of macroharmonic change increases. While this largely corresponds to the general trends present, the shape does not adequately reflect the listener's experience of the section: given the rate of change, the aural macroharmonic memory is considerably more short-term. For example, there is little reason to assume that RUs 430–434, a stretch of five RUs adhering to the PC set of the A-flat major scale, would sound three times more diatonic than say RUs 444–448, a stretch of the same length in a D-flat major scale PC set. However, the force comparison graph for ±6 shows the former reaching 14.28 Lw and the latter sinking to just 3.96 Lw. (This is a result of the effect the PCs in the neighboring scale stretches have on the data in this excessively large span.) A ±2-span (presented below the ±6 version) is considerably more representative: the diatonic force looks much more mountainous with the magnitude of the individual peaks corresponding to the center of each scale stretch, and those for the particular portions mentioned (i.e. RUs 430–434 and 444–448) now seem more reflective of how they might be perceived.

Much like the previous transition between sections, the movement into the final *Molto vivace* is effected with semitone motion through a dynamic drop from *forte* to *pianississimo*. However, this time the voice-leading is not a mere parallel shift. In fact, the final chord of the *Allegro vivace* (consisting of PCs Db, D# and Bb) is the only simultaneity in the section to violate diatonic principles: there is no diatonic scale to which it could belong (hence the final split in the scale motion graph in Fig. 3.25, marking the two diatonic scales this RU is a composite of). The flat notes resolve down to C and A, the sharp up to E, a scalar forking that in fact sets up the processes involved in the *Molto vivace*. Importantly, this first (accented) chord of the

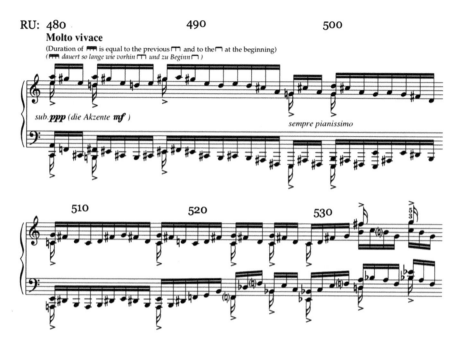

Fig. 3.26 Etude 16, RUs 480–539, score. © 2005 Schott Music, Mainz—Germany

new section is the first, clearly discernible triad of the etude, placed in relief with no obstructing notes.[2]

This final section of the etude resembles the penultimate one in its repetition of a musical unit through rapidly shifting scale areas. While this was a particular verticality in the *Allegro vivace*—namely the IC5-based chord—in the *Molto vivace* it is a set of intervals spread horizontally. These intervals in fact constitute a kind of major hemitonic pentatonic scale. Although in prime form they form the set 01378, the particular arrangement of the PCs in register and dynamics (i.e. the punctuating accents) often creates a somewhat Lydian impression (as can be observed in Fig. 3.26), so this could possibly imply a 0267E arrangement with a root of 0. However, for the purposes of labeling this scale it will be taken as a major-scale subset, implying a stepwise pattern of 0457E for a tonic of 0 (namely the "C major" version).

The *Molto vivace* departs significantly from the previous section in the relationship between the two hands: whereas throughout the *Allegro vivace* they both occupied the same macroharmony, the aforementioned forking at RU 407 leads to a divergence in the scale areas that each hand outlines. For the most part, the shapes of the swirling figurations are coordinated in the two hands (so that they move in

---

[2]There is one clean triad before this point—at RU 360, see Fig. 3.19—but given that it is sandwiched between two accented RUs, the unique nature of the chord would not be evident to the listener.

parallel). However, at certain transition points between different scale areas these shapes diverge in direction, yielding a changing intervallic relationship between the hands. The centrality of IC 5 is still present in the perfect-fifth accented dyads, but the framework in which they function is considerably less strict than in the *Allegro vivace*.

Thus, in order to chart the scalar motion of the *Molto vivace*, the hands need to be separated, as shown in Fig. 3.27 (where the graph is placed between separate Co5 PC graphs for each hand). The scalar motion graph demonstrates a quicker and often erratic motion in the LH, compared to a considerably smoother one in the RH. Both hands show a preference for the sharp key areas, with less time spent in flat keys—a logical result of the broad sharpening taking place through the course of the etude. However, as shown by the rolls, several of the enormous leaps in the scalar motion graph would not be perceived as such: for example, the LH leap of eight upward Co5 steps from −3 (E-flat major) to 5 (B major) at RUs 546–547 would likely be heard as a considerably smaller movement of four steps down (namely to C-flat major).

The shape of the figurations would seem to imply a strong tritone force: the 0267 shape (when transposed to 0) that is constantly outlined is a neat subset of the maximal PC-set for balance 2 (i.e. 012678). Yet the force graph for the two hands together (Fig. 3.28) only shows one main tritone surge peaking at RU 564, even though nothing in the texture in the individual hands would imply such a pattern—the intervallic composition of the figuration has remained the same. It is rather the relationship *between* the two hands that yields a focus on the PC set 012678 (incidentally also the prime form for the maximal PC-set for balance 2). However, given the clear (albeit constantly shifting) bi-tonality of the passage, a listener would not necessarily perceive the macroharmony as a sum of the two hands: rather, there would be two distinct, superimposed scale areas, a feature reinforced by the registral independence (and, especially at the outset of the section, great distance) between the hands. Similarly, the summed correlation strength graphs show a very weak correlation, failing to reflect the strict adherence to the scale type in each scale stretch *within* each hand. As in the *Allegro vivace*, the integrity of the (generally partial) diatonic sets are fully maintained in each scale stretch (especially in the RH, where some stretches of adherence to a single scale type are as long as 34 RUs)—this is a very different process from the general fuzzification of the B-flat minor scale in the first two sections of the etude. Thus, it is considerably more useful to separate the data for the two hands as shown in Figs. 3.29 and 3.30.

The individual-hand force graphs show that the tritone force is indeed the overall strongest force (though not by a large margin), and that it is in the foreground largely throughout the *Molto vivace* rather than surging in just one stretch as it did in Fig. 3.28. Similarly, the correlation strength is far stronger in these graphs than in the summed one, particularly in the RH where the scalar motion is relatively slower (and thus the scale stretches are longer). Interestingly, the strong presence of the tritone force seems to relate back to the *Andante* in an almost cyclic manner—as if the shared variables that connect adjacent sections also link the outer ones.

The macroharmonic independence of the hands is also a continuation of a process that was put in motion in the *Allegro con moto*. While this was somewhat reset by the

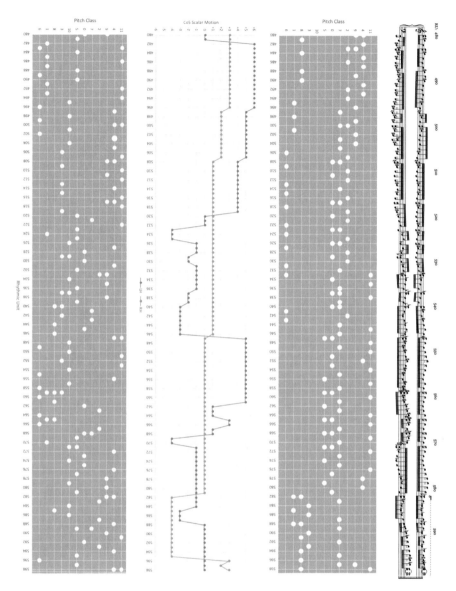

Fig. 3.27 Etude 16, RUs 480–598, comparison of three graph types (RH Co5 PC graph, scalar motion, LH Co5 PC graph)

Allegro vivace, by the end of the section the contrary registral motion and final scalar forking at RU 479 served as a re-articulation of this trend of divergence.

After this extensive cycling through different key areas in the *Molto vivace*, the etude comes to a halt at a final four-note simultaneity—an E-major triad (Fig. 3.31). As mentioned in the discussion of the *Andante*, the E natural was one of the few

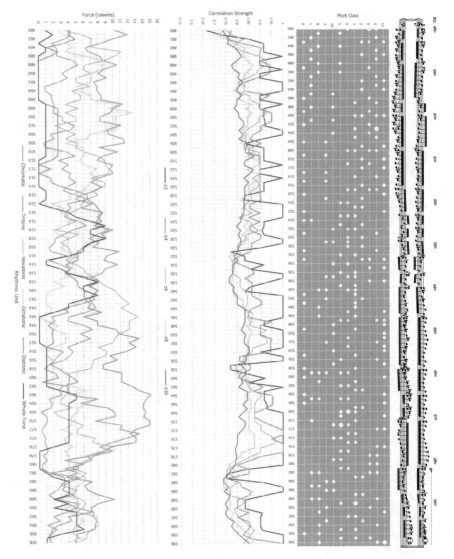

Fig. 3.28 Etude 16, RUs 480–598, comparison of three graph types (Co5 PC graph, 7-PC correlation strength and force comparison for ±6)

foreign notes at the conclusion of that section, and the only one not to belong to the B-flat minor scale area (broadly conceived)—as if a herald of its importance later on in the etude. The arrival on the E-major triad is significant in itself: it is a verification of the final arrival in sharp key areas, and this general sharpening is also reflected in the movement from the opening minor key area to a conclusion in the major. However, it is also part of a larger motion articulated by the *Molto vivace*, given that the section opens with a pivotal A minor triad; it is as if it eventually resolves

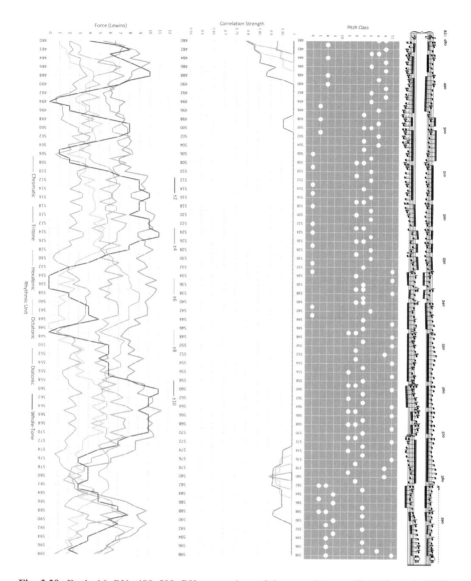

Fig. 3.29 Etude 16, RUs 480–598, RH, comparison of three graph types (Co5 PC graph, 7-PC correlation strength and force comparison for ±6)

into the E major triad. In fact, given their inversion (the A minor triad in first, the E major in second) the chords could comfortably form adjacent steps in a harmonic progression (though not in the particular registral distribution of the chords in the section)—which, with E major as the tonic, could form a kind of minor IV-I progression.

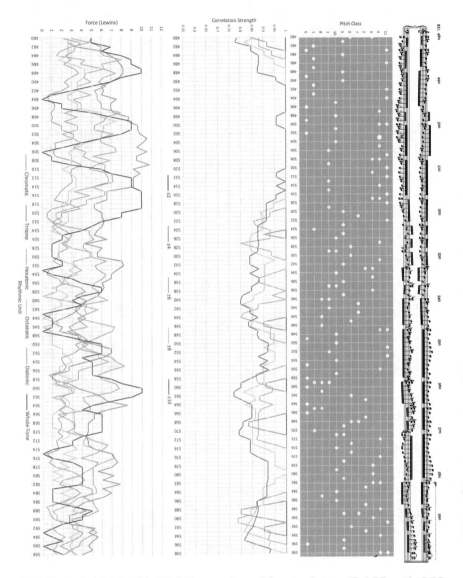

Fig. 3.30 Etude 16, RUs 480–598, LH, comparison of three graph types (Co5 PC graph, 7-PC correlation strength and force comparison for ±6)

However, one could also imagine these triadic beacons as relating even further back in fact to the previous etude, the conclusion of which was anchored on an A minor triad—an initial "tonic" which the opening of the *Molto vivace* in *Pour Irina* then picks up, then shifting to an eventual E major "dominant." In fact, the idea of not concluding on the tonic but rather on a kind of hanging dominant chord that requires resolution (especially given its second inversion) seems very much

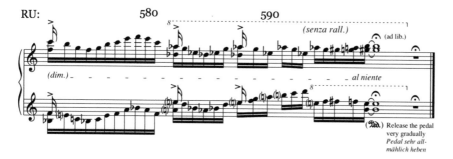

Fig. 3.31 Etude 16, RUs 571–598, score. © 2005 Schott Music, Mainz—Germany

appropriate for *Pour Irina*, where there is no return home in the dramatic narrative: once it sets out from a very particular tonal space it hardly looks back, arriving at an entirely different harmonic area after an increasingly disorienting meander.

Thus this final chord seems to beckon a continuation in the ensuing etudes in the set, where a resolution may be found. While this kind of global harmonic progression across the etudes may seem somewhat far-fetched (and certainly not perceptible to the listener unless they hear the etudes as a set, and even then, would likely not be picked up on without prior knowledge) the harmonic relation of these triadic textural outliers is striking—and will be even more so when we continue to keep an eye (and ear) on this conspicuous pattern in the etudes that follow.

Chapter 4
Articulating Ternary Structures in Etude 17, *À bout de souffle*

4.1 General Remarks

Etude 17, *À bout de souffle* ("Breathless") is the only etude of Book 3 not to contain any tempo changes: the entire work is one mad dash, *Presto con bravura*. However, this is not to say that it does not have a clearly defined structure. In fact, the nature of the etude's formal design is among its most fascinating and surprising features.

The breathlessness of the work is not only due to the quick tempo, but also to the fact that most of the etude is in canon: one hand is constantly in pursuit the other, one or two RUs behind and displaced by an octave. The majority of the motion in the eighth-note *perpetuum mobile* texture is stepwise and in a single voice, with the occasional leap and accented chordal punctuation (Fig. 4.1)—giving the etude a distinct running, *courante*-like quality.

These small twists, turns and constant direction changes in the passagework are reflected by the broader zig-zagging up and down almost the entire keyboard range in the work (see Fig. 4.2); together they create a chase-like effect, where one hand is constantly trying its best to throw the other hand off its trail. The texture eventually screeches to a halt with five chords in increasing duration (a written-out ritardando constituting a dotted quarter note, a dotted half note, a dotted breve, and a dotted breve with a fermata), the pursuing hand finally caught up and the two now in unison—the etude has run out of breath.

These features make the etude an apt reflection of another work of the same title, Jean-Luc Godard's iconic *nouvelle vague* film, in which a young criminal is constantly on the run from the police who remain one step behind until the very end, when they finally catch up and shoot him. In the famous final scene, the wounded criminal staggers, slows down, runs out of steam and drops dead. As with *White on White*, we do not know if the allusion of the title to another artwork of the same name is deliberate; in the case of *À bout de souffle*, the parallels in dramatic narrative likely suggest a certain level of influence.

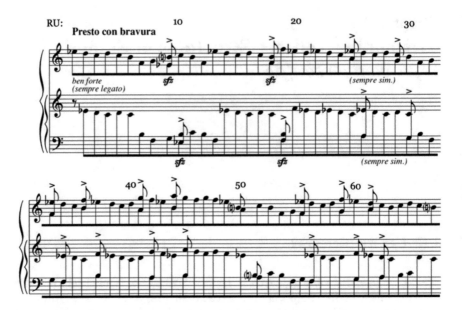

Fig. 4.1 Etude 17, RUs 1–67, score. © 2005 Schott Music, Mainz—Germany

There are two moments where the broader, steady registral motion is broken—at RUs 641 and 812, which frame a central section that is somewhat different from the rest of the passagework. At RU 638 the texture has reached the upper end of the keyboard's treble when the LH (so far in pursuit of the RH, one eighth-note behind and an octave below) suddenly gets cut off for two RUs, reappearing at RU 641 four octaves lower. The RH promptly joins the LH in the lower register, now the pursuer and two eighth-notes behind, yet soon goes off on a tangent, abandoning both the canon and eighth-note motion and instead presenting a leisurely melody in a series of dyads, while the LH continues the running eighth-notes in a quiet muttering in the background (marked "only a shadow" in the score). The RH soon returns to the eighth-note motion and canon (now ahead again), but it is not quite a *perpetuum mobile*, as it constantly starts and stops. Eventually coming to a rest at RU 799, the RH reappears in the middle of the keyboard's range at RU 803—with the same series of pitches with which the etude opened—and the LH, having itself come to a quick stop, leaps up to rejoin the RH in the original configuration. (See Fig. 4.2 for a visual representation of these moments.)

This central section that contrasts with the outer passages creates a kind of ternary, ABA structure (even though the three sections are not quite proportional, with the first being around four times the length of the other two). This form is articulated not only by registral and textural changes, but also by thematic material— namely, the return of the opening "melody." The theme reappears throughout the work at key moments, and a lot of the running passagework is generated from

Fig. 4.2 Etude
17, structural overview—
annotated piano roll (notes
in P-space, C3 = middle C;
green is for foundational
notes, red for those that
correlate to the main key
areas in the structure,
marked in yellow)

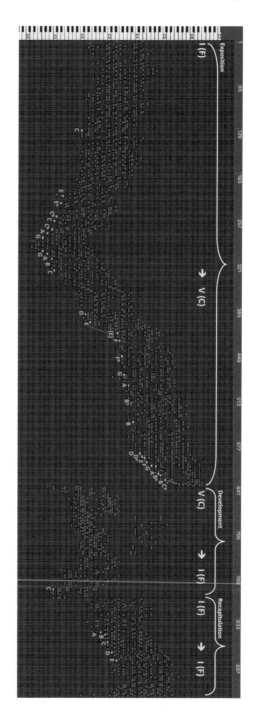

smaller stretches and gestures from that opening shape—Ligeti's fractal-inspired take on motivic development.[1]

The primarily stepwise-moving textures articulate different scales. What is surprising, however, is the manner in which certain scale types and areas dominate, as well as the systematic nature of their juxtaposition as relating to the broader structure. Figure 4.1 shows the opening of the work, which adheres entirely to a rather unusual type of scale—the series {0, 2, 3, 5, 7, 9, 11} grounded on 5. While the proportion of tone- to semitone- steps in this 7-PC set is that of the diatonic, the ordering is not. This is in fact the "acoustic" scale, namely the major scale with a raised fourth and flattened seventh degree (thus with the first accidental in both sharp and flat directions).

The acoustic scale is sometimes called the ascending melodic minor—which in this case is somewhat misleading, as for the above PC-set the latter would be grounded on C, whereas in the opening of Etude 17 the F is clearly the first degree of the scale. The name "Lydian dominant" is slightly better, implying a Lydian scale with a flattened (thus dominant) seventh degree. "Overtone" scale is also used, given the raised fourth and flattened seventh degrees that appear in the first 11 steps of the harmonic series (adjusted to equal temperament). This paper, however, shall use the term acoustic, as that is the name for this scale type in Bartók scholarship—and in Hungary in general—and is thus almost certainly the term Ligeti would have used for it.

The acoustic scale not only opens the work but is also the primary scale type towards or against which motion occurs—unlike in the other etudes of Book 3, where it is the diatonic (and its subsets) that serves this function. (In Etude 17, the diatonic scale generally functions as an alteration of the acoustic.) Another respect in which *À bout de souffle* is unique among these last etudes is the primary significance of pitch space for the listener's interpretation of macroharmonic quality (rather than it being mostly determined by PC space). Given the steady movement up and down the keyboard, certain pitches take on a grounding role, appearing repeatedly as the lower bounds of the texture. This registral arrangement translates to PC space, where these notes function as a kind of first degree or foundation on which the rest of the PCs are built—in the manner that the F does in Fig. 4.1. However, these foundational notes do not always correspond to the first degree of the scale type: a

---

[1] It is important to clarify what is here considered the theme or melody. The term applies to the shape that is outlined by the upper notes of any stretch of this shape—thus, at RU 9 in the RH the B natural would be considered the melody note, not the Eb below. The confusing factor here is that it is the *lower* note that is connected to the beam of the rest of the melody notes. This is due to the fact that Ligeti beams all RUs of the running eighth-notes together below the staff in both hands, with the accented dyads always having an upward stem—see Fig. 4.1. (RUs 641–802 are an exception due to textural changes to be addressed later on.) Ligeti always connects the lower notes of the dyads to the global bottom beam and thus avoids framing the notes of the dyads with stems on either side—this is simply a question of visual arrangement and does not have significance in terms of voice allocation. The only time Ligeti diverts from this system is when the dyads are seconds, meaning that the notes of the dyads are not vertically superimposed upon one another, but rather displaced horizontally; this allows Ligeti to occasionally tie the upper notes to the lower beam (see, for example, the RH at RU 336 in Fig. 4.23).

texture could belong entirely to the Bb acoustic scale yet with the C functioning as the foundational note.[2]

Thus, the treatment of scale in this etude has a distinctly modal quality. The shifts in the dominant PC-sets do not necessarily coincide with changes in foundational notes, causing a constant aural reevaluation of what constitutes the tonic and thus compounding the sense of motion and instability. This makes the moments where they do coincide—especially for extended passages—particularly important, and Ligeti reserves such occurrences for key moments in the work.

The central claim of this chapter is that Ligeti uses the juxtaposition of macroharmony and thematic material to articulate a ternary structure that is essentially a kind of sonata form. As bizarre as this may sound, the etude in fact charts a I-V-I path, with a transition from I to V in the exposition, a contrasting development section that begins rooted on V and then modulates back, and a recapitulation that returns to the opening material and continues with a transition in the manner of the exposition but finally returns to the tonic key area.

This opening theme in fact appears at a number of important moments of this structure beyond just the beginning of the recapitulation; it serves as a marker of certain key arrivals rather than functioning in opposition to a certain second subject (as understood in the post-Czerny formulations of sonata form). In addition, the choice of foundational notes, the duration for which they are active as well as the correlation strength of the texture to the primary scale types all serve to underline these landing spots of the form.

Figure 4.2—an annotated, color-coded piano roll—gives a broad outline of the structure of the work. The commentary following the graphical analysis will elucidate how the rest of the visual representations (namely the PC rolls, the correlation strength and force graphs) further highlight this formal design.

It is worth noting that the question of which note in a texture constitutes a foundational note is not always clear-cut. In the 67 RUs of Fig. 4.1, for instance, the F is not the lowest note, as there is an Eb below it. However, the F is far more frequent and is repeatedly present, while the Eb appears only once—making the F a better candidate for the foundational note. Passages throughout the rest of the etude are more ambiguous. The foundational notes marked in green in Fig. 4.2 are derived from a combination of several factors, such as cardinality, accentuation, registration and position relative to the foundational notes before and after it—both horizontally (namely in terms of the number of RUs that elapse between them) and vertically (for example, in a broadly upward motion, a foundational note is generally not considered supplanted by a lower note appearing after it).

---

[2]In such circumstances, the "acoustic" term has less of an advantage over, for instance, "melodic minor," given the modal treatment. For the above example (C, D, E, F, G, Ab, Bb) this book would use the term "C acoustic Dorian" (as a result of considering the Ionian/major scale the "home base" for the diatonic scale type).

Before proceeding, it is necessary to discuss in some detail the question of interpreting the force graphs for a scale type that does not correspond to their representative PC-sets.[3] In the rest of the etudes of Book 3, the primary scale type—namely the diatonic scale and its subsets—are altogether adequately reflected by the diatonic force. While other scale types (such as the whole-tone or octatonic) are less frequently used, they have their counterparts among the Fourier Balances. In *À bout de souffle*, however, the primary scale type is the acoustic, which, despite similarities to the diatonic, does not readily correspond to any of the force types.[4] The following discussion will focus on the acoustic scale, though principles from it can be derived for any scale that is not emblematic of the six force types.

In order to gain an understanding on how the acoustic scale functions on the Fourier Balances, one can start out by simply lining up the acoustic scale to each of the maximum-force inducing PC-sets—henceforth called "maximal" sets—to identify any particular similarities (Fig. 4.3). It is clear that the acoustic scale has the most in common with the maximal sets for balances 4–6. Given that in terms of the ratio of scalar tone to semitone steps the acoustic has most in common with the diatonic—and that Ligeti often uses these two scale types as alternates of one another in Etude 17—it is also worth lining up these two scales as shown in Fig. 4.4.[5]

The next step would be to simply plot the acoustic scale on each of the balances, then record and compare the resulting forces. Any acoustic scale will yield 0.73 Lw on balance 1, 0 Lw on balance 2, 1 Lw on balance 3, 2 Lw on balance 4, 2.73 on balance 5 and 3 Lw on balance 6. At a first glance, balance 6 (for whole-tone force) seems the most representative of the acoustic scale. This however initially seems at odds with the idea that the acoustic is most closely related to the diatonic, given the step-ratio described above. Furthermore, the acoustic scale has the same ratio of PCs that belong to the maximal sets vs. their complement for balance 5 as it does for balance 6—namely five common PCs and two counterbalancing ones (Fig. 4.3).

This discrepancy is explained by comparing the force the acoustic scale yields on each balance to the force that the maximal set for each would cause, as these in fact vary from balance to balance. This comparison is shown in Table 4.1, which also includes the diatonic scale for reference.

Table 4.1 shows that the resulting force from the acoustic scale is somewhat weaker than that of the diatonic scale on balance 5 (namely 73% as strong), identical on balance 3, double on balance 4 and in the range of triple for balances 1 and

---

[3] For a detailed discussion on what PC-sets may be considered scales, see Carey (2002). The forces from balances 4, 5 and 6 (and to some extent 3) can comfortably be considered reflective of scale types; those for balances 1 and 2 are less clear-cut.

[4] The PCs on balance 5 (or any of them for that matter) cannot be simply reordered to correspond to a synthetic scale such as the acoustic without violating the basic principles and functionality of the balance.

[5] The ratio is of five tone to two semitone steps in a total of seven steps. The diatonic major pattern is T-T-S-T-T-T-S, while the acoustic is T-T-T-S-T-S-T.

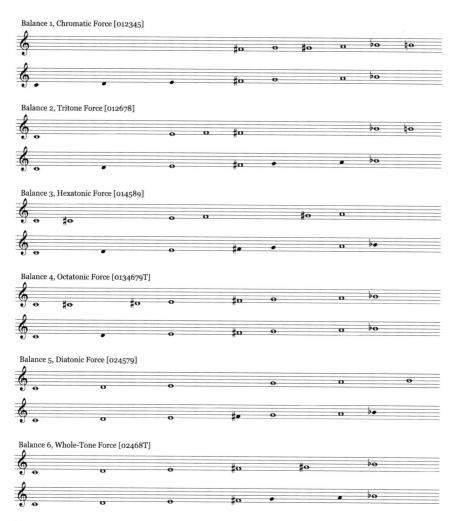

Fig. 4.3 Comparison of the acoustic scale to maximal sets for each balance (the C acoustic scale is placed below the maximal set for each; common PCs are lined up vertically, and conflicting notes in the acoustic scale marked with black instead of white note-heads)

Fig. 4.4 Comparison of the acoustic scale to the diatonic (C acoustic scale below C Lydian; common PCs are lined up vertically, and conflicting notes in the acoustic scale marked with black instead of white note-heads)

Table 4.1 Comparisons of forces (in Lw) from three PC collections on each balance (the diatonic scale, acoustic scale and maximal sets)

	Maximal set		Diatonic scale		Acoustic scale		
Balance	Set	Force	Force	Percentage of max.	Force	Percentage of max.	Percentage change (dia:ac)
1	[012345]	3.86	0.27	7	0.73	19	270
2	[012678]	4.00	1.00	25	0.00	0	0
3	[014589]	4.24	1.00	24	1.00	24	100
4	[0134679T]	4.00	1.00	25	2.00	50	200
5	[024579]	3.86	3.73	97	2.73	71	73
6	[02468T]	6.00	1.00	17	3.00	50	300

6. More importantly, it confirms that the acoustic scale is indeed more diatonic than it is whole-tone, as the force it yields on balance 5 is 71% as strong as the maximal set, whereas it is only 50% as strong on balance 6. It results in more overall Lewins on balance 6 is simply because the maximal set on balance 6 causes 6 Lw, compared to 3.86 Lw on balance 5.

In fact, the series of balance numbers 1 through 6 do not seem to line up with the maximal forces at all, which form a peculiar sequence—one that derives from the geometric properties of the Fourier Balances. There is an equation (albeit a somewhat convoluted one) that maps balance numbers onto the maximal forces. In algebraic terms, the maximal force m on any balance b can be expressed as follows:

$$m(b) = \sqrt{\left(\sum_{n}^{n+5} \sin\left(\frac{\pi b}{6}\left(2n - \frac{n(b_{mod6})}{b}\right)\right)\right)^2 + \left(\sum_{n}^{n+5} \cos\left(\frac{\pi b}{6}\left(2n - \frac{n(b_{mod6})}{b}\right)\right)\right)^2}$$

A detailed discussion on how this equation was developed here, its underlying geometric principles and the way in which it may be adapted for use in other contexts can be found in Appendix B. For now it suffices to simply keep in mind the question of maximum possible forces and the way in which readings from the balances reflect a scale (such as the acoustic) that does not fit neatly on any one of them. This will inform the manner in which one interprets the force graphs for Etude 17 and uses them to identify macroharmonies and their quality. However, it is important to note that force readings such as those in Table 4.1 are for sets of notes where there is one of each PC in the scales (or PC sets) described. The practical application here is of course rather different—in the textures analyzed, some PCs appear more frequently than others, and their ratio is what will affect the scalar force results. For example, a passage where the acoustic scale is dominant but its fifth and sixth scale degrees are relatively infrequent will register a considerable whole-tone force; one where the second degree appears less frequently will yield a strong result on balance 4, and so forth.

As mentioned above, in the other three etudes of Book 3 the primary scale type towards or against which motion occurs in the diatonic, so the force readings on balance 5 as compared to the rest are the usual starting point for analyzing macroharmony. The centrality of a different scale type in Etude 17—especially one as atypical as the acoustic—makes the situation rather more complex. The many ways in which the acoustic scale is in fact reflected by the force graphs means that there can be multiple markers of this primary scale type which will need to be taken into account. The particular scalar force that is especially dominant for any passage will be a result of two features: the cardinalities of the various PCs of the scale in the texture, and the difference in maximal forces for each force type (where the whole-tone force has a natural advantage over the rest).

While these are in fact considerations with any reading of any scale type, the ambivalent nature of the acoustic scale makes them especially important. In contrast, the far more clear-cut nature of the diatonic scale—with its unequivocally strong showing on balance 5 compared to the rest—somewhat mitigates the discrepancies caused by the irregularity in maximal forces across the six balances, for example. The fact that the acoustic scale does not have an emblematic force type also means that the correlation strength readings will be especially important in determining macroharmonic stability. Similarly, the shape that the acoustic scale takes on the PC rolls (especially in harmonic orderings) can also be a useful identifier. Thus, the availability of multiple markers for scale correlation reflect the adaptability of this system of graphical analysis for scale and PC-set types that do not readily correspond to those represented by the Fourier Balances—in any PC universe.

4.2 Graphical Analysis (Figs. 4.5, 4.6, 4.7, 4.8, 4.9, 4.10, 4.11, 4.12, 4.13 and 4.14)

Fig. 4.5 Etude 17, RUs
1–943, piano roll (notes in
P-space, C3 = middle C)

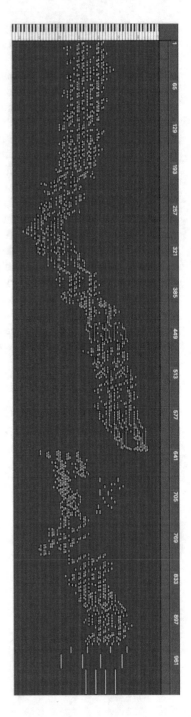

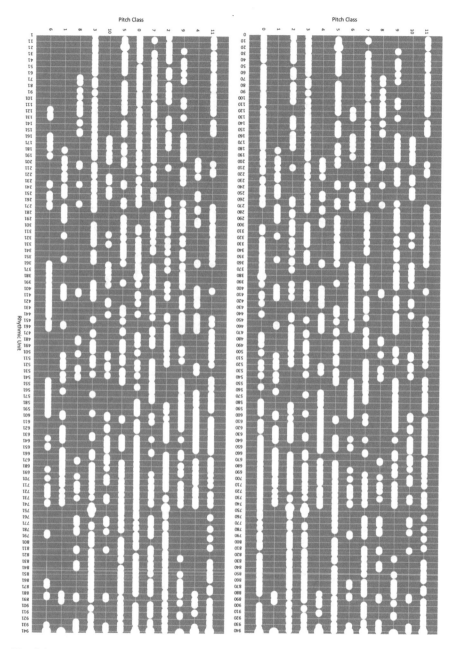

Fig. 4.6 Etude 17, RUs 1-943, PC graphs. (**a**) (top): chromatic (notes placed in PC space ordered chromatically). (**b**) (bottom): Co5 (notes placed in PC space ordered by the Co5)

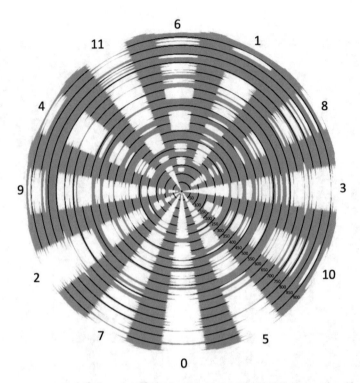

Fig. 4.7 Etude 17, RUs 1–943, conical Co5 PC graph (notes placed in PC space ordered by the Co5 using polar coordinates: the y-axis is rendered as a circle, and the black rings mark the passage along the x axis)

4.3 Commentary

The melodic shape with which the etude begins is significant, given its later reoccurrences. Although hardly tune-like, its motion and particular configuration of steps and leaps has a very distinct profile that makes it easily identifiable—even when varied in most of its reiterations (Fig. 4.15).

The opening of the etude displays a confluence of three key factors—scale, foundational note, and correlation. The foundational note is F throughout, the scale area is F-acoustic, and the texture does not stray from the seven PCs of this set for 67 RUs, which is reflected in the complete correlation strength (see Fig. 4.17; the correlation strength begins to wane at the end due to the first foreign note at RU 68). These features do not always go hand-in-hand in the etude; as mentioned in the introduction of this chapter, the foundational notes and scale area do not often line up, and such strict adherence to the dominant PC set for extended stretches is rare. Thus, the opening of the etude represents relative stability, the home ground for later departures—and any passage that displays a similar level of stability will be especially important to the listener's perception of the structure.

A glance at the conical Co5 PC graph (Fig. 4.16) shows the extent to which the PCs of the F-acoustic scale—namely PCs 0, 2, 4, 5, 7, 9 and 11—are central

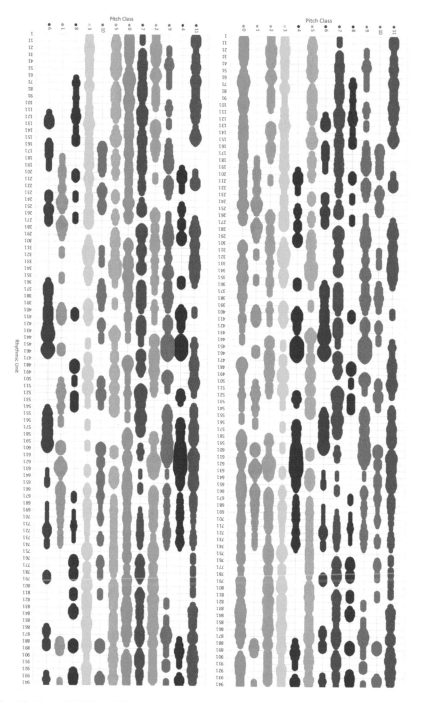

Fig. 4.8 Etude 17, RUs 1-943, PC graphs. (**a**) (top): chromatic PC graph, ±6 span (notes placed in PC space ordered chromatically, with a span of ±6 applied to each RU). (**b**) (bottom): Co5 PC graph, ±6 span (notes placed in PC space ordered by the Co5, with a span of ±6 applied to each RU)

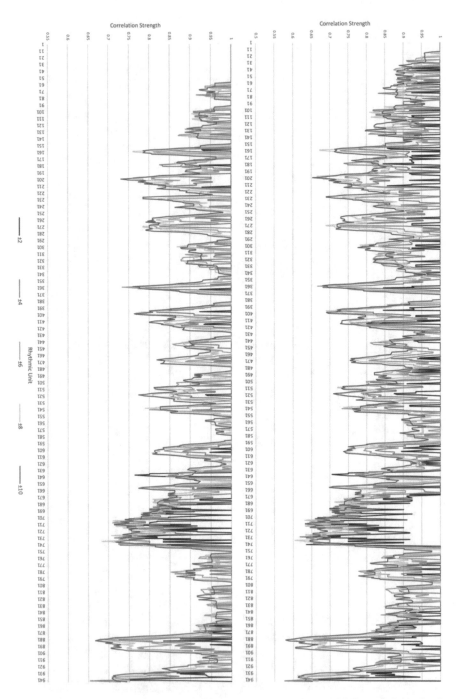

Fig. 4.9 Etude 17, RUs 1-943, PC correlation strength graphs. (**a**) (top): 6-PC correlation strength graph (number of notes that belong to the 6 most frequently appearing PCs/total number of notes). (**b**) (bottom): 7-PC correlation strength (number of notes that belong to the 7 most frequently appearing PCs/total number of notes)

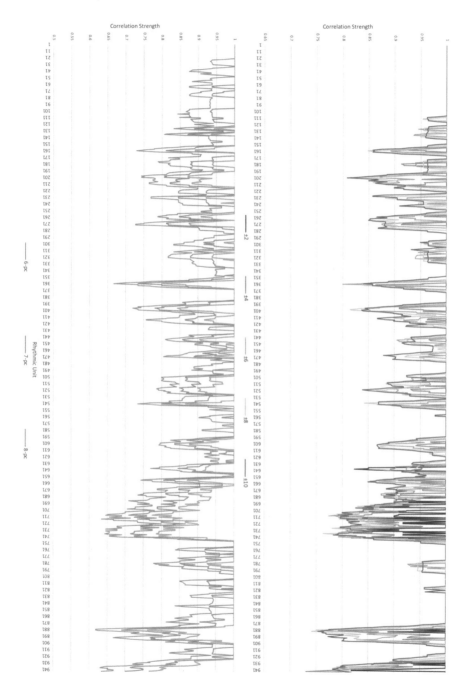

Fig. 4.10 Etude 17, RUs 1-943, PC correlation strength graphs. (**a**) (top): 8-PC correlation strength (number of notes that belong to the 8 most frequently appearing PCs/total number of notes). (**b**) (bottom): correlation strength comparison (comparison across 6, 7 and 8-PC sets using ±6)

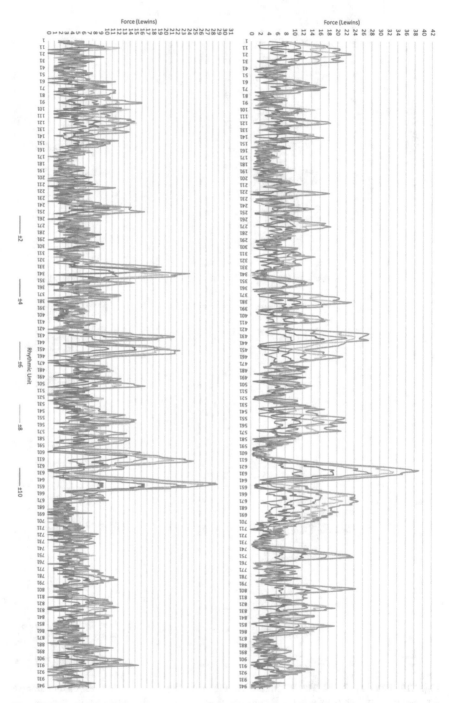

Fig. 4.11 Etude 17, RUs 1-943, force graphs. (**a**) (top): chromatic force graph (force applied to Fourier Balance 1, measured in Lewins—span comparison). (**b**) (bottom): tritone force graph (force applied to Fourier Balance 2, measured in Lewins—span comparison)

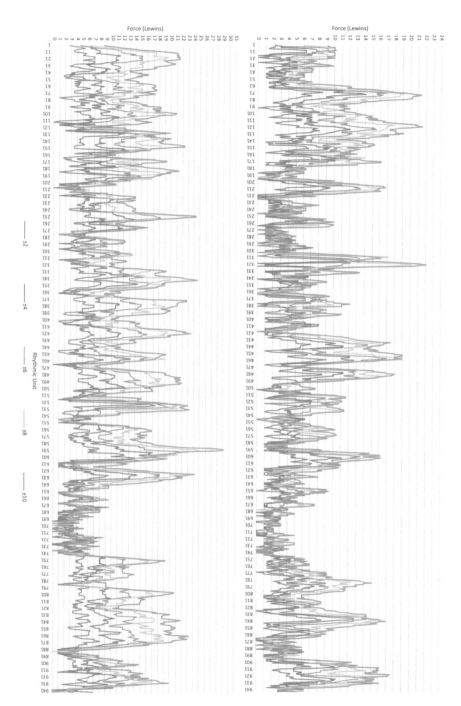

Fig. 4.12 Etude 17, RUs 1-943, force graphs. (**a**) (top): hexatonic force graph (force applied to Fourier Balance 3, measured in Lewins—span comparison). (**b**) (bottom): octatonic force graph (force applied to Fourier Balance 3, measured in Lewins—span comparison)

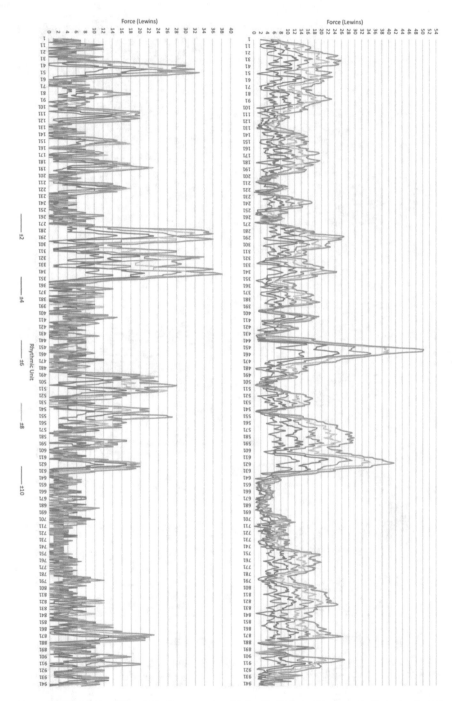

Fig. 4.13 Etude 17, RUs 1-943, force graphs. (**a**) (top): diatonic force graph (force applied to Fourier Balance 5, measured in Lewins—span comparison). (**b**) (bottom): whole-tone force graph (force applied to Fourier Balance 6, measured in Lewins—span comparison)

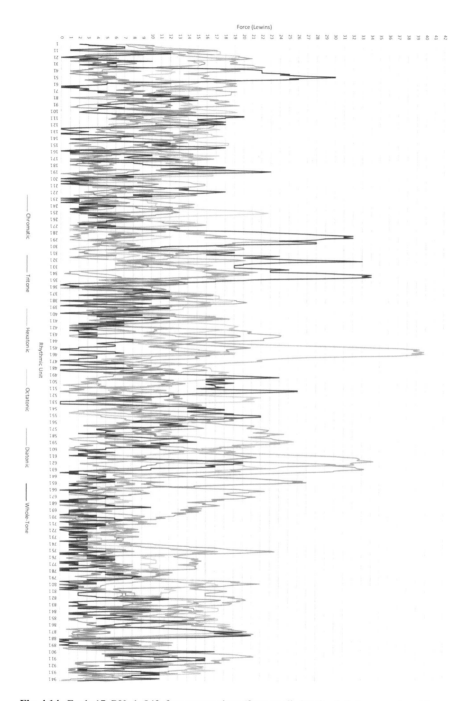

Fig. 4.14 Etude 17, RUs 1–943, force comparison (force applied to the six balances, using ±6 as a representative span)

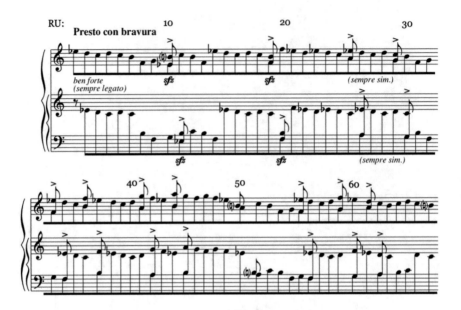

Fig. 4.15 Etude 17, RUs 1–67, score (repeat of Fig. 4.1). © 2005 Schott Music, Mainz—Germany

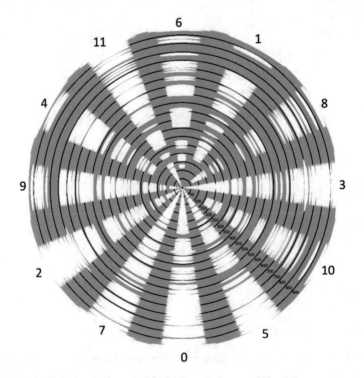

Fig. 4.16 Etude 17, RUs 1–943, conical Co5 PC graph (repeat of Fig. 4.7)

throughout the whole work, being clearly more dominant than PCs 1, 3, 6, 8, and 10. It also displays the particular shape that the acoustic scale takes on the Co5—a central five PCs occupying adjacent steps, flanked by two PCs that are displaced one step out. Relative to the F major scale, the E has been moved out a step to become B, and the Bb out to an Eb. Figures 4.16 and 4.35 visually demonstrate the exclusive presence of this pattern in the first 67 RUs of the work.

It is worth taking a look at the force graph comparison in this opening clean stretch of the acoustic scale to see the kind of readings this scale type registers (Fig. 4.17). As we can expect, the stability of this passage is not reflected in a dominant diatonic force (unlike in the rest of the etudes of Book 3). Instead, different forces come to the fore depending on which PCs of the acoustic scale are particularly present in various stretches. For example, RUs 10–24 focus entirely on PCs 0, 2, 3, 5 and 11 (with no other PCs present), which as a subset of the octatonic scale makes the force for balance 4 stronger than the rest at RU 16 in Fig. 4.8. At RUs 14–26, PCs 0, 2, 3 and 5 are dominant (with 6, 9, 7, and 4 appearances, respectively), with only two other PCs present (9 and 11), each of which appear only twice. This clustering in a narrow band of the gamut causes a surge in the chromatic force (RU 20). The diatonic force does eventually come to the fore at RU 30—all PCs of the acoustic scale are present at RUs 24–36, but PCs 9 and 11 appear least often, and as they are two of the PCs that form tritone relationships in the scale their absence causes less of a canceling out on balance 5. Finally, the overwhelming predominance of the whole-tone scale 1 subset {3, 5, 7, 9, 11} at RUs 42–55 (where the only other PC to appear is 0, turning up only twice) causes a surge in whole-tone force.

The contrast between the total, constant correlation strength and the volatile variance in the force readings of Fig. 4.17—where it seems as if four different force types are jostling for primacy—demonstrates the extent to which the acoustic scale does not fit neatly onto any one balance. The opening of Etude 16, *Pour Irina*, showed how highlighting a certain portion of a scale (such as omitting the sixth degree of the diatonic major scale and giving the tritone relationship particular prominence) can yield strong force readings for balances that are not emblematic of the full set (which in the case of Etude 16 was balance 2 for tritone force); in *À bout de souffle* this is even more pronounced, with shifting stresses on particular PCs of the acoustic scale causing alternations in the predominant force types. This makes the correlation strength graphs particularly important representations in determining levels of macroharmonic stability in Etude 17.

The first PC to undermine the opening acoustic set is an Ab at RU 68—the first step down on the Co5 beyond the boundaries of the acoustic.[6] Ligeti continues this familiar process with the introduction of F# (the first step up) as the next foreign note at RU 118; even though the acoustic scale is not reflective of the Co5 in the manner the diatonic is, principles of movement in harmonic space are still present. (This also hints at the fact that diatonic scales will indeed play an important role in the etude

---

[6]Given the canon, when discussing musical events the references to RUs are always for the RH, unless explicitly noted otherwise.

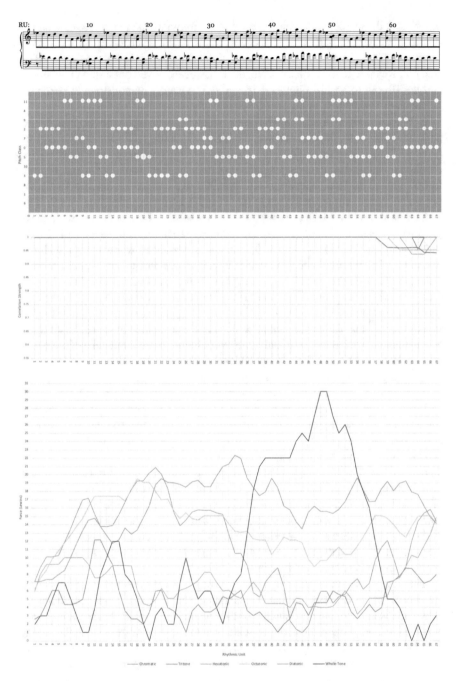

Fig. 4.17 Etude 17, RUs 1–67, comparison of three graph types (Co5 PC graph, 7-PC correlation strength and force comparison for ±6)

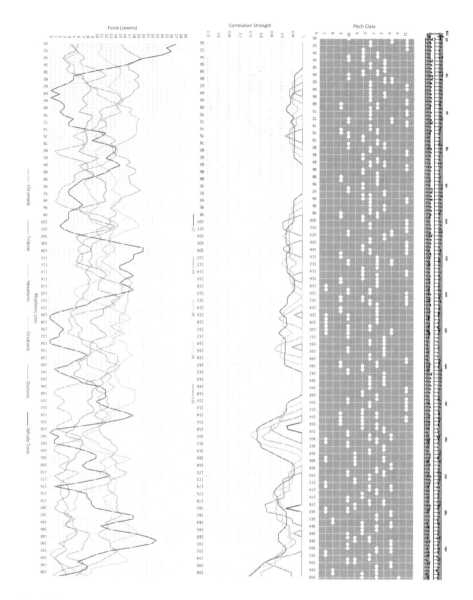

Fig. 4.18 Etude 17, RUs 50–199, comparison of three graph types (Co5 PC graph, 7-PC correlation strength and force comparison for ±6)

further on.) However, this widening of the gamut does not cause a modulation from the dominant macroharmony per se—the new additions do not replace their natural counterparts, as A and F continue to still be present. Thus the opening macroharmony is not supplanted but gets fuzzier, reflected in the steadily decreasing correlation strength (Fig. 4.18). By RU 198, all remaining PCs have been used, and

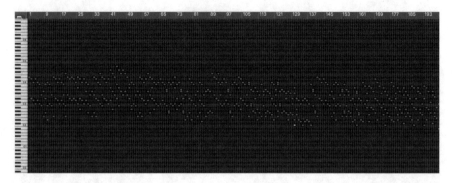

Fig. 4.19 Etude 17, RUs 1–200, color-coded piano roll (notes in P-space, C3 = middle C; green is for foundational notes; red for those that correlate to the main key areas in the structure)

by this point the correlation strength is at one of its weakest points in the entire etude. This process is also reflected by a general weakening in the scalar forces (Fig. 4.18). Despite occasional spikes caused by certain combinations of PCs (such as the introduction of F# causing a strong octatonic force in RUs 126–138—see RU 132), the forces eventually peter out by RU 200 due to the chromatic saturation.

By this point, the F has been supplanted by C as the foundational note (Fig. 4.19); by moving to a scale degree that, in common-practice tonality, is a point of tension that beckons a resolution to the tonic, the texture seems to underline the inherent instability of the fuzzification. This moment also sets up the larger movement between I and V that forms the trajectory of the etude.

The final foreign note to be introduced is the E natural at RU 198—which is significant, as it is an alteration (and hence refutation) of the PC with which the etude began (namely Eb). This E functions as a kind of boundary for the opening, given that it causes a change in the motion and a final breaking out from the opening scale area. It first induces descending chromatic lines—which are not only a novelty in terms of motion type, but also dissipate the dominance of any seven-PC scale area; this chromatic moment contains the first respellings in the etude (Fb for E, Db for C#), heralding a directional change. Then at RU 214 the passagework leaps below the foundational note C (in terms of register in each hand)—and starts exploring new scale areas.

At RU 227 the opening theme returns for a stretch of 24 RUs. However, it does not occupy one single scale area and instead alternates between the Eb acoustic and Dorian scales.[7] The shape of the theme is also slightly altered, and the difference is something that could be measured by what is here called "thematic correlation." The intervals between each consecutive note pair (of the upper line, ignoring the lower notes in the accented dyads) in the new iteration are compared to the original in two ways—generic (whether it is a second or fourth step for example) and specific (for

---

[7]While it could also be considered a Db major scale, the presence of Eb as a kind of foundational note implies the Dorian.

Fig. 4.20 Etude 17, RUs 227–250, thematic correlation to RUs 1–24

particular ICs, such as IC 1, 6 etc.), adopting the terminology of scale theory developed by John Clough (see Clough and Myerson 1985, 249), in the manner shown in Fig. 4.20. A value is assigned for the difference in each of these spaces: a second becoming a fourth is a difference of two, IC 2 being replaced by IC 5 is a difference of three, and so forth. The difference for each space can be a maximum of three for the generic and six for the specific, as the separating intervals are considered in prime form for both; the total correlation ranges from 1 (where all intervals are identical) to 0 (where all are the furthest distance apart). Any temporally inserted or subtracted notes are assigned the maximum values for both generic and specific intervals (as shown further on in Fig. 4.25).

Figure 4.20 shows a high degree of both generic and specific thematic correlation; the stronger showing for the generic reflects the fact that the shape of the theme has been kept, but the shifting scale areas have resulted in chromatic alterations.

Following this reiteration of the theme the texture begins to plunge down the keyboard; the wavelike shape of RUs 233–238 takes on a motivic function, repeating several times on progressively lower foundational notes. This fast hurtling through different scale areas causes a number of interesting effects, such as a quick cycling down the Co5 (given the progressively flatter scale areas) and the fact that smaller scalar force spans in fact overtake large ones, which suffer from encompassing multiple scale areas—until the texture hits upon the G acoustic scale at RU 277 (first on foundational note A, but then on pitches below) which lasts longer and causes a surge in the whole-tone force in particular given the focus on PCs 1, 5, 7, 9 and 11, as shown in Fig. 4.21.

The G acoustic scale yields to the original macroharmony, F acoustic, coordinated at RU 300 with a landing on F as the foundational note. Despite a few chromatic alterations (reminiscent of the modal processes at RUs 227–250) the F acoustic scale maintains its dominance through RU 358; during this passage, the foundational notes creep up an octave and come to a rest on F again, which is repeated as the lower bounds of the texture no fewer than eight times during RUs 333–358. This sense of regaining scalar stability is reflected by the steadily rising correlation strength and the surge in whole-tone force (the strongest marker of the acoustic scale), illustrated in Fig. 4.22. However, there is an absence of not only the opening theme, but any material that could be identified as thematic; instead, the accented dyads increase in number, creating a sense of angularity and agitation by often appearing on consecutive RUs. Despite the reaffirmation of the

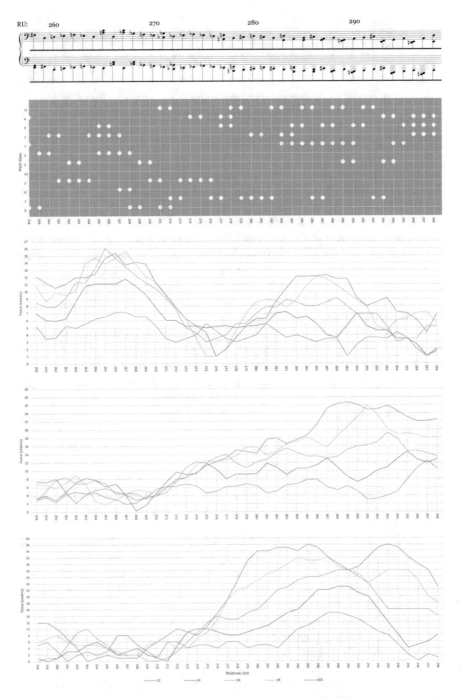

Fig. 4.21 Etude 17, RUs 258–298, comparison of four graph types (Co5 PC graph and octatonic, diatonic and whole-tone forces)

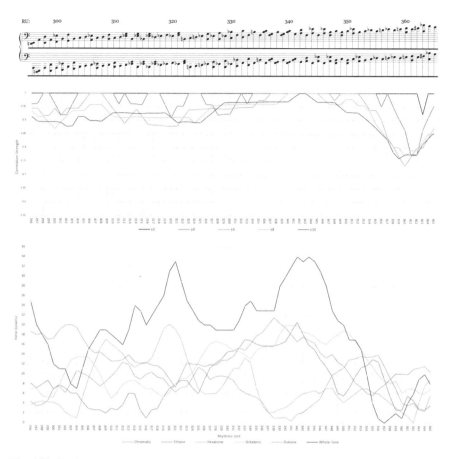

Fig. 4.22 Etude 17, RUs 296–365, comparison of three graph types (7-PC correlation strength and force comparison for ±6)

original macroharmony, tension is built in anticipation of an impending shift. The PC to open the floodgates, so to speak, is the very same one that did so the first time around (at RU 198)—E natural. Promptly after its appearance at RU 359 the texture suddenly shifts to a new macroharmony (and foundational note), sending both the correlation strength and scalar forces plummeting (Fig. 4.22).

The plunge in these markers is due to the sudden juxtaposition of these scale areas, instead of a steady transition between them. However, the constitution of the new macroharmony is not dissimilar to the previous one: it is another acoustic scale, now on C, and the foundational note has also shifted up to C. This I-V motion is further underlined by an exact repetition, starting at RU 363, of the first 24 RUs of the opening (a 100% generic and specific thematic correlation), now transposed to C (Fig. 4.23). Even the positions and intervals of the accented dyads (beyond the first RU) are exactly as in the opening theme.

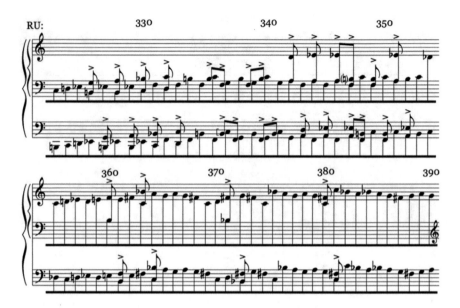

Fig. 4.23 Etude 17, RUs 323–390, score. © 2005 Schott Music, Mainz—Germany

With this confluence of scale, foundational note and total thematic correlation, the passage from RU 363 creates a level of stability greater than at any stretch after the opening of the work—a clear structural marker that establishes the fifth degree of the tonic F acoustic scale as the second key area. The significance of the I-V relationship was already evident from the extended presence of C as the foundational note at RUs 130–213, and the reappearance of the opening theme cleanly transposed to the dominant key confirms this beyond any doubt. The uniqueness of this moment is further highlighted by the 7-PC correlation strength, which promptly recovers after the rapid macroharmony switch and reaches full correlation for the longest stretch since the beginning of the work (and, in fact, for the rest of the work as well)—Ligeti has evidently cleared the air for this new scale area to establish itself (Fig. 4.24).

A destabilization follows, with similarities to the procedures in the first key area. The first foreign note is Eb at RU 395—which has the same relationship to the C acoustic scale as Ab (the first foreign note in the opening) had to F, and the foundational notes continue to chart a relatively scalar path (save for an octave displacement at RUs 405–421—see Fig. 4.24).

However, the foundational notes now steadily rise instead of initially descending as in the first key area—and they begin their journey promptly after the reiteration of the opening theme, rather than waiting out the steady fuzzification process at the beginning of the work; as soon as the texture begins to shift upwards in register it also starts cycling through a number of different scale areas, some of which yield especially strong force readings. These are due to both the nature of the PC-sets highlighted as well as to the considerably denser textures. For example, at RUs

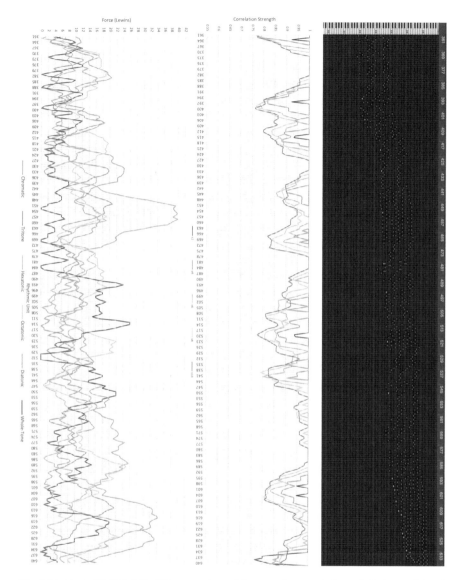

Fig. 4.24 Etude 17, RUs 363–640, comparison of three graph types (color-coded piano roll, 7-PC correlation strength and force comparison for ±6)

443–465 a series of accented chords on every other RU creates a kind of hocketing between the hands, causing a near-doubling of the number of notes in the 23-RU stretch. Given the fact that the center of the passage highlights a diatonic 024579 hexachord through the dominance of PCs 1, 2, 4, 6, 9 and 11 (only one other PC appears—PC 7—and only once), the diatonic force surges (Fig. 4.24).

In a manner similar to the first part of the exposition (namely that of the tonic key area), the dominant has its own modified iteration of the opening theme, starting at RU 537 and lasting for 39 RUs—the most extended return of the opening theme. It is first set up by 12 preceding RUs that establish the macroharmony—an alternation between the Bb acoustic and Ab major scales, thus the same acoustic/Dorian exchange as at the modified appearance of the theme at RU 227. However, here the foundational note is not Bb, but in fact C—further affirming the presence of the second key area. It also results in the acoustic scale being itself treated modally, becoming a "C acoustic Dorian" scale (i.e. 0, 2, 4, 5, 7, 8, 10), juxtaposed with a C Phrygian, the grounding minor second of which (namely between its first and second scale degrees) creates an inherent instability. This volatility seeps into the onset of the theme at RU 537 which promptly respells the Db as a C#, causing an eventual shift to the A acoustic scale (though vestiges of the C Phrygian/acoustic Dorian areas persist for a while) and a loss of a clear foundational note. The thematic correlation itself remains strong (despite the adding of an extra RU within the shape), registering 0.97 for the generic and 0.92 for the specific (Fig. 4.25).

As the passagework nears the upper boundary of the keyboard, the registral range of the texture narrows—across the two hands as well as within each one—as if getting squeezed in. This concentration is also reflected by the frequency of the accents: for the first time in the etude, more than two consecutive RUs within each hand are accented, now for as many as 11 straight RUs, creating an increase in dynamic volume. For a stretch of 28 RUs (at RUs 607–634) an E natural seems to be the upper boundary, until an F is finally reached at RU 635—both significant PCs, one the pivotal PC that initiated the modulations out of the tonic scale areas, the other tonic itself. This narrowing texture briefly outlines the white-note set, in particular the set {0, 2, 4, 7, 9, E} which causes another surge in diatonic force (peaking at RU 615), while the final packing of all PCs within the range of a tritone (while hitting up against the F boundary) results in a strong chromatic force (Fig. 4.24).

The process of crashing against a boundary at the far end of the keyboard's range is one that frequently appears in Ligeti's etudes, primarily at the end of a section (or the entire work itself). The deceleration in registral motion that leads to this moment creates the expectation of some form of caesura. At RU 637 the LH abandons the canon, shoots up and crashes against the RH—quite literally, as its motion is halted entirely for two RUs, the first rests in the etude (since the opening eighth-note displacement of the LH). The RH itself is cut off at RU 640; the LH reappears at the next RU five octaves down, now *pianissimo* and ahead of the RH,

Fig. 4.25 Etude 17, RUs 537–575, thematic correlation to RUs 1–38 (the inserted note is assigned the maximum values for both generic and specific intervals)

which then follows along in canon two RUs late. This theatrical moment and first break in the texture clearly indicates the onset of a new section in the work (Fig. 4.26).

However, the opening theme (the primary marker of important structural moments thus far) is nowhere to be found. In fact, the texture hardly resembles that of the previous scalar passagework at all, with an angular texture outlining parallel tritones making way for a chromatic slithering—where all the dyads, now unaccented, are also tritones (resulting in a predictable surge in both tritone and chromatic forces, as shown by Fig. 4.27). This change in texture after the break at RU 640 makes clear that this new section is of a fundamentally different constitution than the preceding passages, and one that is considerably more ambivalent with regard to any scalar or tonal anchoring: a development after the previous exposition. Interestingly, the structural elements that resemble sonata form in the etude are emphasized by the fact that this development section lifts off from C as the clear foundational note, thus articulating the onset of the V-I motion that forms the function of this section.

The central passage of the development (RUs 682–743) is the greatest textural anomaly in *À bout de souffle*, further highlighting the extent to which the development departs from the framing exposition-recapitulation sections. These 103 RUs also pose something of a challenge for the graphs, given the fact that the textures for

Fig. 4.26 Etude 17, RUs 618–675, score. © 2005 Schott Music, Mainz—Germany

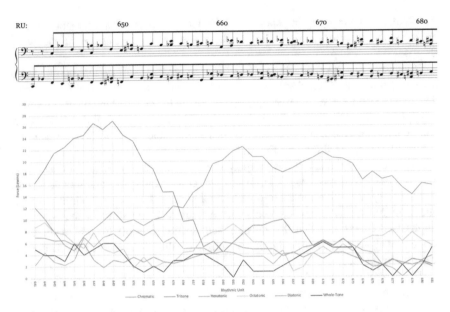

Fig. 4.27 Etude 17, RUs 641–681, force comparison

the two hands diverge almost entirely—the listener is conscious of hearing two distinct channels, the features of which would be neutralized by any representation that collects all of the data together.

At RU 681, the RH breaks off for two RUs. The LH continues with the hushed running eighth-note motion, now always in two voices—but as with the rest of the development, none of these dyads are accented. All of the intervals in this LH texture are either major or minor seconds/sevenths (namely ICs 1 and 2) or tritones (ICs 6), which, with the chromatic meandering, continue the "anti-tonal" nature of the section (Fig. 4.28).

The reappearance of the RH at RU 684 violates several of the principles that governed the relationship of the two hands thus far, such as the canon, registral separation by an octave, eighth-note motion and matching dynamic levels. Instead, it appears two octaves above the LH, *mezzo-forte* (marked in *rilievo*, "in relief"), and presents a melodic shape in two voices, *cantabile*. The RH texture in fact seems in direct opposition to what is happening simultaneously in the LH. Not only does it abandon the *perpetuum mobile* to instead move in far larger RUs and depart from the muted dynamics, but it also refutes the chromatic saturation of the LH, instead outlining none other than the F acoustic scale (with only one outlier, the final Bb).

Each note of this shape has an outsized presence compared to the underlying regular eighth-notes. This is due to their greater duration, as well as the fact the relative to the LH, the RH is in the clear foreground in terms of register and dynamics. However, these are features that are not picked up by the graphs (save for the piano roll, which represents pitch space). Instead, the sheer volume of notes in the LH overwhelms the data, so that the representations that further process the PC

Fig. 4.28 Etude 17, RUs 675–743, score. © 2005 Schott Music, Mainz—Germany

data—namely the correlation and force graphs—end up being largely reflections of the LH texture (Fig. 4.29).

When produced in this manner, the graphs fail to represent what a listener may experience here—which is either the presence of two superimposed and competing layers, or in fact the primacy of the layer that outlines the F acoustic scale. (The extent to which a listener will interpret the RH as more dominant compared to the LH will depend somewhat on the performance, acoustics and so forth.) Figure 4.31 instead shows a clearly dominant chromatic force (for the first portion of this passage where the LH occupies a narrower gamut in both pitch and PC) followed by a weak showing for all forces (the most extended such moment in the whole work, and in fact in any etude of Book 3) due to the chromatic saturation.

There are a number of possible workarounds to this problem—none of which are a perfect solution but may approach a more accurate representation of the listener's experience. The first is to simply separate the two hands, in order to better reflect the distinct nature of these two layers (Fig. 4.30). However, the forces in the LH are a

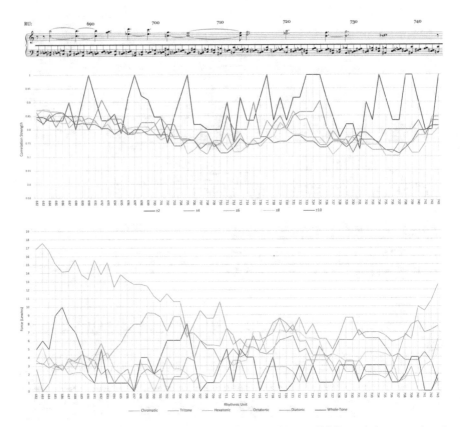

Fig. 4.29 Etude 17, RUs 682–743, comparison of two graph types (7-PC correlation strength and force comparison for ±6)

good deal stronger—even during the chromatic saturation—given the set cardinalities, as the RH notes are only recorded at the point of attack.

In order to compensate for this, one could repeat the attack of each RH note on each RU for the length of their duration. For example, for the first RH dyad at RU 684, the PC data 0, 1 would be entered for each of the nine RUs 684–692. While this has the advantage of not ignoring note length, it is generally a greater distortion of the data as any note on the piano will decrease in volume soon after being struck, becoming less present with the passage of time. Thus, it will only be comparable to another note of the same dynamic level at the point of attack. However, in the current circumstances this is somewhat less of an issue as the RH is in fact meant to be a good deal louder than the LH. The failure of the graphs to reflect dynamics is here somewhat offset by the ballooning of RH data through the note repetitions, which while still distorting the nature of the note decay can at least reflect the primacy of the RH in this texture. Figure 4.31 shows the separate-RH force graph with note repetitions, followed by the summing graph for the two hands using this data

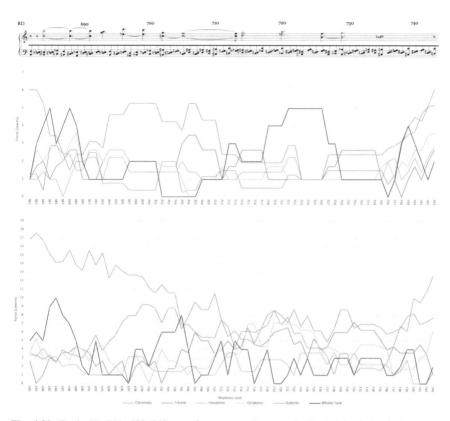

Fig. 4.30 Etude 17, RUs 682–743, two force-comparison graphs for ±6 (each hand alone, RH above LH)

collection method. Figure 4.32 compares the correlation strength for the RH separately and the two together, using the note-repetition method.

Figure 4.31 shows the extent to which the repeated-note method for longer notes brings the RH material to the foreground, both in the much stronger force showings for the RH alone as well as the way these results are reflected in the summation of both hands (rather than the RH data causing little more than a slight adjustment to the otherwise LH-dominated total data, as in Fig. 4.30); a similar trend can be seen in the correlation strengths in Fig. 4.32 (especially when compared to the weaker correlation force in Fig. 4.29). For the scalar force comparison, the types most clearly associated with the acoustic—the diatonic and whole-tone—are clearly dominant, with the former peaking at RU 708 (given the adjacent Co5 steps 0, 5 and 7 in the RH) and the latter at RU 727 (due to the whole-tone {3, 5, 7, 9, 11} portion of the acoustic scale).

At RU 744 the texture snaps back to the original configuration (the RH ahead of the LH by one RU) with elaborations on portions of the opening theme. However, the RH seems to have trouble picking up the momentum it lost in the previous

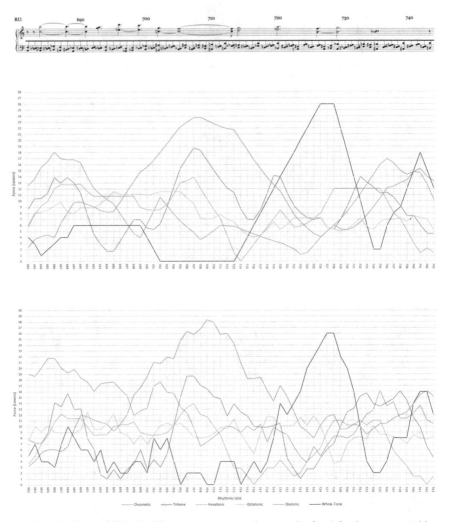

Fig. 4.31 Etude 17, RUs 682–743, two force comparison graphs for ±6 using note repetition (RH alone followed by the two hands together)

stretch, occasionally coming to rest on longer notes and thus creating more melodic, phrased utterances, as if somewhere halfway between the driven passagework in the exposition and the relaxed interlude at RUs 682–743. It begins with an alternation between the F acoustic and Dorian scales (a pairing already used a number of times in the etude, such as at RUs 227–250) and soon introduces, within the acoustic passages, the same two foreign notes Ab and F# first heard at the outset of the work (Fig. 4.33).

This is a remarkable manner of effecting the V-I motion in the development. Ligeti sets up these two areas as opposing dichotomies on the far ends of this section: the first (RUs 641–681) chromatically saturated and disorienting, with no melodic

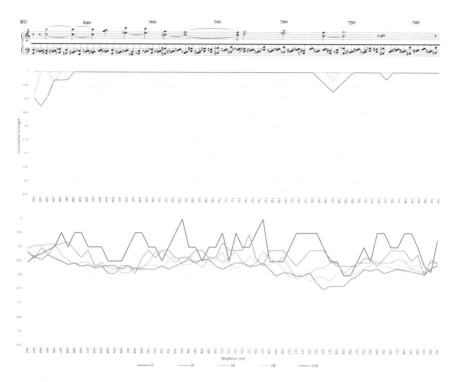

Fig. 4.32 Etude 17, RUs 682–743, two correlation strength graphs using note repetition (RH alone followed by the two hands together)

shape or scalar correlation yet anchored on C, the second (RUs 744–802) with a clear theme and macroharmonic profile built on F. Then, instead of modulating from one to the other he simply creates something of a cross-fade in the middle for RUs 682–743, with the LH maintaining the features of the preceding section and the RH passage anticipating those of the following one. Revealingly, the LH passagework at RUs 682–743 keeps C as the foundational note, further evidencing how this texture correlates to what came before; once the RH tentatively returns to the canon, F is reestablished as the foundational note (Fig. 4.34). A brief descent follows, circling again around C but now in the context of the F acoustic scale and thus setting up the dominant-tonic motion that is achieved with the return of the opening theme at RU 803. For the first time in the etude, the theme returns in its original pitch space (Fig. 4.33)—a clear marker of the recapitulation.

Given that the recapitulation is a total of 141 RUs compared to the 640 RUs of the exposition, the processes are considerably truncated. The shape of the opening theme is strictly followed through only 8 RUs only before a Dorian coloring; the same fuzzification process as at the opening, with the introduction of Ab and then F#, begins only 32 RUs into the recapitulation (instead of the 68 RUs for the exposition). The foundational notes steadily rise and soon lead to strings of accented

Fig. 4.33 Etude 17, RUs 744–832, score. © 2005 Schott Music, Mainz—Germany

dyads on consecutive RUs, in a manner similar to the agitated passages at the registral peak of the exposition (RUs 595–605, for example)—the recapitulation has reached these textures of maximum tension in less than 100 RUs, whereas in the exposition it took about six times as long.

After this rapid destabilization the texture lands at RU 896 on foundational note F and subsequently the F acoustic scale: the return to the tonic key area after the transition (in place of the dominant as in the exposition). There is only one PC that violates this scale, the pivotal E natural; it is as if this final return to F contains a summation of both the anchoring and undermining elements that defined the dramatic narrative of the etude. (This anomalous PC will in fact prove to have even great significance further on.) The presence of the E natural within the otherwise strictly F-acoustic context colors the passage by the white-note set (arguably the

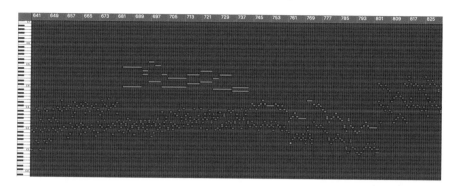

Fig. 4.34 Etude 17, RUs 641–832, color-coded piano roll (notes in P-space, C3 = middle C; green is for foundational notes; red for those that correlate to the main key areas in the structure)

primary blank slate in these final four etudes) and causes a rise in the diatonic force (peaking at RU 907). Though significant relative to the other peaks in scalar force in this section, the recapitulation cycles through key areas too rapidly for any one of them to truly take hold (Fig. 4.35).

Ever since the start of the recapitulation the dynamics have been slowly increasing (starting at *mezzo-piano*, in contrast to the exposition which was *ben forte* throughout). This crescendo highlights the density and tension of this final section, and by the return of the tonic key area the dynamics are approaching *fortissimo*. The Ab and F# appear again at RUs 926 and 931, a final undermining of the acoustic scale that leads to its demise: the passage reaches a double *fortissimo* (*ffff*) before a sudden, almost cinematic cut to a different atmosphere at RU 938: *subito pianississimo*, with a series of progressively longer RUs. Crucially, the LH in fact catches up with the RH 940, and the rest of the RUs have the two hands finally coordinated (Fig. 4.36). At the risk of overstating the cinematic parallels in Etude 17, the last *ffff* chord comes across as something of a gunshot—at long last, the pursuit is over.

The first unison RU (namely RU 940) is also the only triad in the whole work–an E-minor one, built on the PC that has played such an important role throughout the work. In fact, the E natural is the only PC to appear twice in these final five longer chords (ignoring simultaneous octave doublings) which otherwise complete the chromatic gamut of 12 PCs, giving this PC added significance.

Thus, this moment of rhythmic consonance—where after 939 RUs the hands finally play "together"—is highlighted by acoustic consonance, one that sticks out dramatically in an otherwise exceedingly dissonant work (where the primary simultaneity is the second, as a result of a canon-displacement for stepwise, scalar motion). While the etude eventually comes to rest on an Eb-A tritone (a simultaneity emblematic of the F acoustic scale), the memory of the minor triad lingers on as a rare moment of rest and the first point of real rhythmic and acoustic resolution in the work.

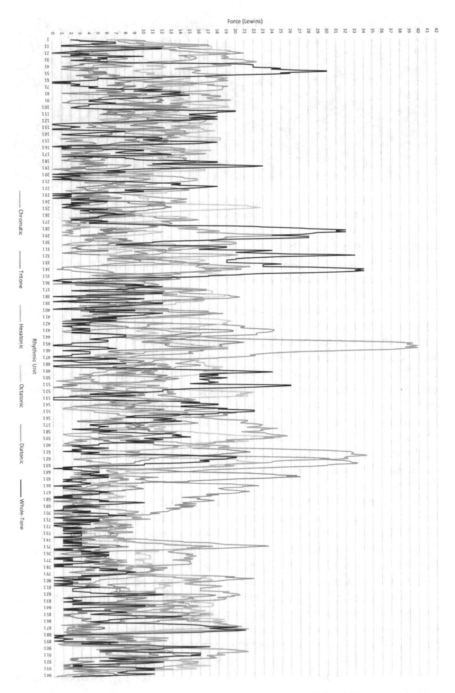

Fig. 4.35 Etude 17, RUs 1–943, force comparison for ±6 (repeat of Fig. 4.14)

Fig. 4.36 Etude 17, RUs 926–943, score. © 2005 Schott Music, Mainz—Germany

What is especially interesting is that this triad fits within a broader pattern that the four etudes of Book 3 seem to chart. The conclusion of Etude 15 is anchored on the A minor triad; Etude 16 ends with a section that articulates a movement from an A minor triad to an E major one; and here in Etude 17, the E minor triad announces the conclusion of the work. In each of the etudes, these triads are outliers in the texture, coming across as moments of particular importance to the listener. While *À bout de souffle* charts a I-V-I path, it is as if the four etudes seem to be forming a certain meta-progression for which the triads of A and E are clear acoustic beacons, themselves forming a I-V relationship. For the eventual resolution of this path we will have to turn to Ligeti's final etude, *Canon*.

Chapter 5
Tonal Procedures in Etude 18, *Canon*

5.1 General Remarks

Ligeti's groundbreaking final piano etude is in two main sections. The bulk of the work is a *Vivace poco rubato* (Fig. 5.1) in unbroken running eighth notes that is then repeated in full even faster, *Prestissimo*—interestingly, the first appearance of the repeat symbol in all of his piano etudes.[1] This is followed by a brief epilogue, *Lento con tenerezza* (Fig. 5.2), comprised of a string of trichords in each hand arriving at a final A-minor triad. Both sections are in canon: in the *Vivace*, the LH follows the RH two octaves lower with a delay of two RUs, whereas in the conclusion the LH is an octave down with a delay of a single RU. In the *Vivace* there are two voices in each hand (save for occasional single notes, which may be interpreted as unisons), making *Canon* is something of a double-note etude.[2] This thick yet extremely rapid texture poses such extreme technical challenges that Ligeti originally considered calling the piece *Casse doigt*—"Fingerbreaker" (Steinitz 2003, 164).

This analysis seeks to show that *Canon* is shaped by an alternation between various largely diatonic scale areas. At the outset, Ligeti establishes the white-note set as the home territory; a number of modulations take place to shift the texture to other diatonic areas, before a movement back to the white-note set. With the exception of one, the very last of them, these modulations take place with a mostly linear, deliberate motion around the Co5: movements away from the white note set move downwards (i.e. in the flat direction), and those going back up. There are three such "zigzags" that take place—away and back, down and up—that connect the four touchdowns in the home key. The *Lento* constitutes the final "zag," the only modulation to use a procedure other than Co5 motion.

---

[1] This section will henceforth be referred to simply as the *Vivace*.

[2] In this paper, unison refers to note doublings within each hand as well as octave PC doublings across the two hands.

N. Namoradze, *Ligeti's Macroharmonies*, Computational Music Science,
https://doi.org/10.1007/978-3-030-85694-6_5

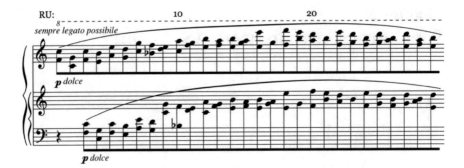

Fig. 5.1 Etude 18, RUs 1–29, score. © 2005 Schott Music, Mainz—Germany

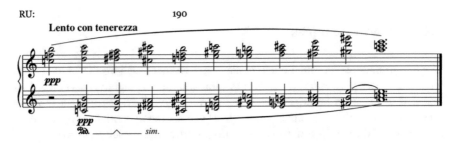

Fig. 5.2 Etude 18, RUs 186–196, score. © 2005 Schott Music, Mainz—Germany

Furthermore, there is a clear palindromic structure to the bulk of the work, encompassing the first two zigzags. The center of this palindrome is what is here called the narrowed or reduced section. These terms refer to the modulations and scale areas, respectively: the modulations are more streamlined, occupying a narrower span around the Co5 compared to the outer, wider motions, and the three central scale areas are not full diatonic sets, but rather are based on transpositions of the Guidonian hexachord, 024579. The final zigzag is an outlier not simply because it does not fit within this palindrome, but because the procedures employed in this section—such as the nature of the modulations or scale area arrivals—depart in important ways from those in the rest of the etude.

This structure is outlined in Fig. 5.3, which uses a modified version of a Co5 PC graph: there are vertical extensions so that certain PCs appear on both ends of the y-axis to better illustrate the function they play at the extreme ends of the various modulations. (The standard version of the Co5 PC graph can be found in Fig. 5.5.) Below the PC graph is a breakdown of the harmonic motion, describing the scale area arrivals as well as the modulation types. Roman numerals indicate stages in the progress of the etude and are used for reference between the graphs and descriptions. Dashed double lines mark the separation of the main palindrome from the final zigzag, dashed single lines keep track of the motion of the texture, and solid curved lines mark symmetrical pairs on either side of the line of symmetry. Given the rather different nature of the procedures in the *Lento*, the graphs and corresponding

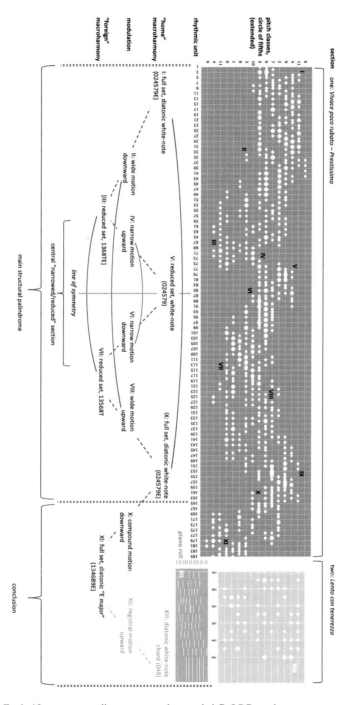

Fig. 5.3 Etude 18, structure outline—annotated, extended Co5 PC graph

descriptions are in a lighter color tone, and a piano roll representation is included below the PC graph to illustrate points made in the analysis below.

Beyond features marked in the structural breakdown, Ligeti deploys additional textural variables to further color the various macroharmony areas. Certain kinds of intervals, motion and voice-leading tendencies are associated with types of PC sets, creating textural profiles that serve to highlight the contrasts between the scale areas. These not only give the macroharmonies their own unique character and sonority but also exaggerate their differences, making the diatonic (especially white-note) scale areas sound even "cleaner" and more consonant, and the chromatic diversions dissonant and murkier. Instead of the alternations between dissonant diatonicity and consonant chromaticism with which Ligeti experiments in his previous etudes (such as *White on White* or *Pour Irina*), here he creates a stark dichotomy of stable diatonic consonance versus unstable dissonant chromaticism. These features, especially when combined with the unusually consonant, triadic final chord of the work (particularly considering a somewhat functional arrival to it, as will be explored below) and the deliberate use of the Co5 for modulations make *Canon* his arguably most "tonal" etude.

Two other features affect the listener's perception of the scale areas. Stark dynamic contrasts in the form of occasional loud accents highlight certain PCs; these often serve to solidify the scale area the texture has arrived at. Although there are no sudden changes in registration within each section (the *Vivace*, save for an initial upward gesture, is one long downward spiral, whereas the *Lento* is a brief ascent), the local registration, namely how the PCs are arranged vertically relative to one another within the texture, sometimes affects our understanding of the macroharmony; this occurs for example by the establishing of certain registrally distinct bass notes that come to the fore. In a work where so many features of the overall sonority have been pared down (the homogeneous rhythm, strict canon and constant two-voice texture for example) the features that *do* vary become especially important in creating contrast and interest for a listener, as well as being particularly revealing for the analyst in understanding the construction of the etude.

The bird's-eye view of the work shown in Fig. 5.3 begins to reveal an implicit dramatic narrative. A diatonic scale area is established in a lengthy opening passage—and not just any diatonic macroharmony at that, but the white-note set, the "cleanest" such scale. The purity of this passage begins to be undermined by chromatic intrusions that provoke a slide both down the keyboard and the Co5. The spiraling down the piano's registers is inexorable, and while there are repeated attempts to shift back up in harmonic motion to the white-note set, any return is short-lived and promptly undermined by new chromatic intrusions—an inability to escape from a pre-destined fate, headed to the abyss. The texture gets increasingly agitated and eventually crashes at the bottom of the keyboard, having failed to return to home ground. This leads to another, more impatient attempt at the whole exercise, now *Prestissimo*.

The majority of Ligeti's etudes end at one or both of the keyboard's registral extremes, either fading away or in a forceful crash—and we seem to be heading towards the latter. However, on the other side of the *Prestissimo* lies a sudden, disorienting stasis: the *Lento*, marked *con tenerezza* ("with tenderness") and *pianississimo*. Ten slow chords, as if suspended in time and space, gently climb to a final, A-minor triad. Not only is this the moment we finally return to and come to rest at home in this etude—it is arguably the only true, final arrival in all of his etudes.

For a composer keenly aware of the implications of his artistic choices on the appraisal of his legacy, especially when it came to his oft-pondered question of "late style" as discussed in the introductory chapter, this manner of ending a piece—a radical departure from his earlier work—is highly significant; in what is not only his last piano etude but his final work it is even more so. While one cannot be certain of Ligeti's awareness of this work as his swan song, his health was already deteriorating when he was writing *Canon*, compelling him to effectively retire after its completion. This chapter will conclude with a discussion of the implications of the *Lento* as the last utterance of Ligeti's oeuvre.

The graphical analyses in *Canon* separate the etude into two distinct sections (the *Vivace* and ensuing *Lento*) and present them in turn rather than together. The transition between them is so stark (in register, tempo, dynamics) that a continuous graph would distort the data; the movement in the *Lento* is so slow that equating its RUs alongside those in the preceding *Vivace* on one x-axis would be of little analytical use. Indeed, the *Lento* is far from the type of flowing, *perpetuum mobile* texture that forms the basis for macroharmony analysis in this book, rendering the use of spans that chart macroharmonic motion over long stretches superfluous. Hence, the only visual representations used for the *Lento* are the ones that display PCs and their scalar forces taken one RU at a time, rather than over larger sets, meaning that representations that use spans (such as the individual force graphs) are excluded from graphical analyses of the *Lento*.

5.2   Graphical Analysis (Figs. 5.4, 5.5, 5.6, 5.7, 5.8, 5.9, 5.10, 5.11, 5.12, 5.13, 5.14, 5.15, 5.16 and 5.17)

Fig. 5.4 Etude 18, RUs
1–185, piano roll (notes in
P-space, C3 = middle C)

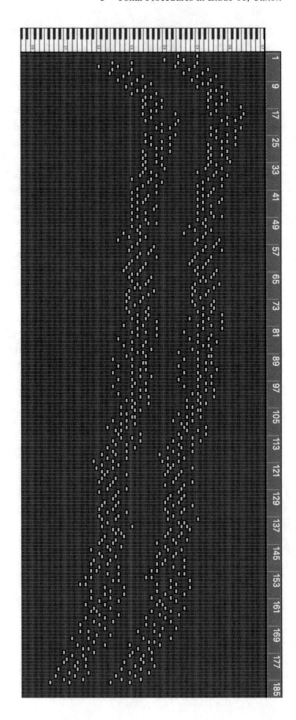

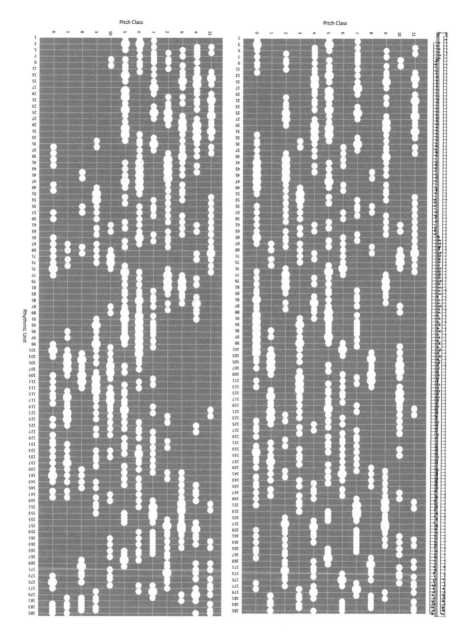

Fig. 5.5 Etude 18, RUs 1-185, PC graphs. (**a**) (top): chromatic PC graph (notes placed in PC space ordered chromatically). (**b**) (bottom): Co5 PC graph (notes placed in PC space ordered by the Co5)

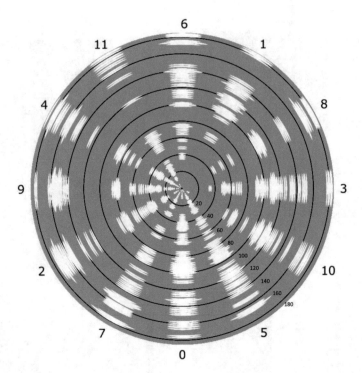

Fig. 5.6 Etude 18, RUs 1–185, PCs in Co5, conical Co5 PC graph (notes placed in PC space ordered by the Co5 using polar coordinates: the y-axis is rendered as a circle, and the black rings mark the passage along the x axis)

5.3   Commentary

Etude 18 is framed not only by diatonicity but by extreme consonance: the work opens with a string of seven RUs employing only white notes, all of the dyads in each hand are perfect fifths, and crucially, Ligeti calibrates the motion in the texture so that the first simultaneity involving both hands actually constitutes a perfect fifth of F and C in unison, two octaves apart (Fig. 5.18). This will be reflected by the last simultaneity of the work, namely the two hands playing a root-position A-minor triad in unison (an octave apart). The narrative of the rest of the work serves to bridge these two pillars of diatonic consonance. The predominance of the perfect fifth in the opening also serves to foreshadow the importance throughout the work of both this interval (including more generally its IC) and the arrangement of PCs across the Co5.

This focus on the perfect fifth in fact brings about the first non-white note: in order to avoid a tritone when the upper voice of the texture reaches F at RU 8 the lower

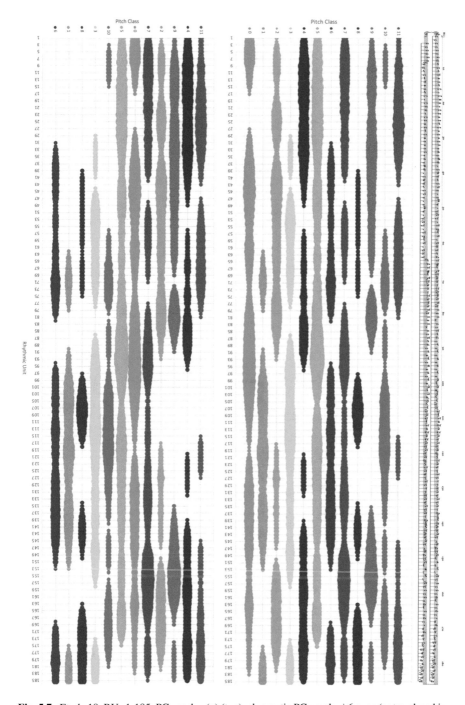

Fig. 5.7 Etude 18, RUs 1-185, PC graphs. (**a**) (top): chromatic PC graph, ±6 span (notes placed in PC space ordered chromatically, with a span of ±6 applied to each RU). (**b**) (bottom): Co5 PC graph, ±6 span (notes placed in PC space ordered by the Co5, with a span of ±6 applied to each RU)

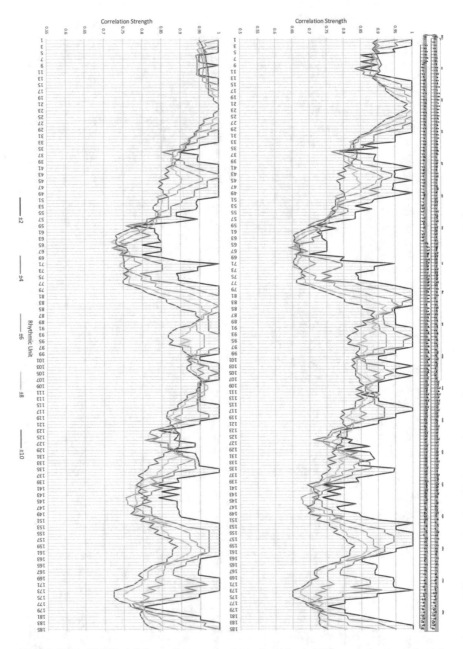

Fig. 5.8 Etude 18, RUs 1-185, correlation strength graph. (**a**) (top): 6-PC correlation strength graph (number of notes that belong to the 6 most frequently appearing PCs/total number of notes). (**b**) (bottom): 7-PC correlation strength (number of notes that belong to the 7 most frequently appearing PCs/total number of notes)

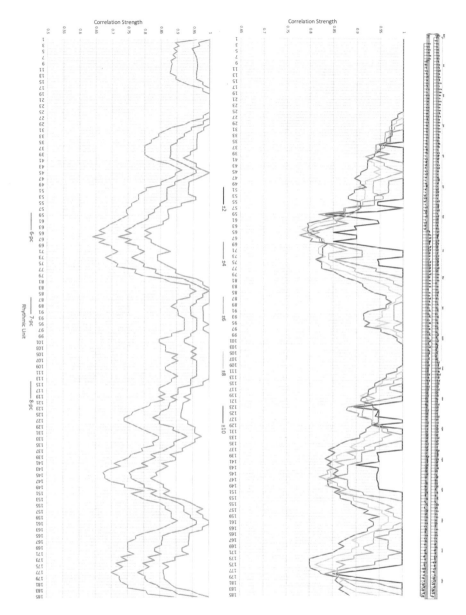

Fig. 5.9 Etude 18, RUs 1-185, correlation strength graph. (**a**) (top): 8-PC correlation strength (number of notes that belong to the 8 most frequently appearing PCs/total number of notes). (**b**) (bottom): correlation strength comparison (comparison across 6, 7 and 8-PC sets using ±6)

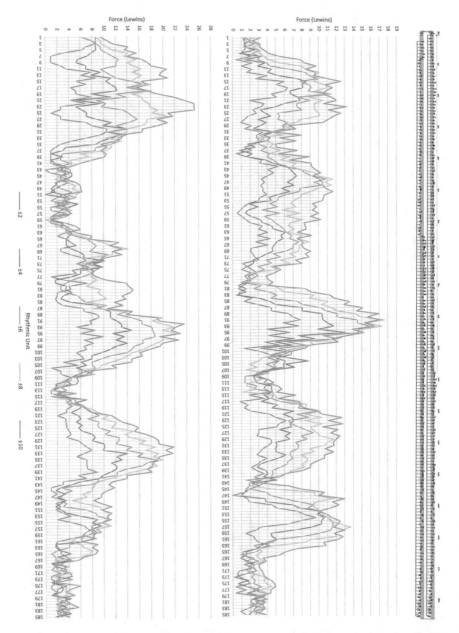

Fig. 5.10 Etude 18, RUs 1-185, force graphs. (**a**) (top): chromatic force graph (force applied to Fourier Balance 1, measured in Lewins—span comparison). (**b**) (bottom): tritone force graph (force applied to Fourier Balance 2, measured in Lewins—span comparison)

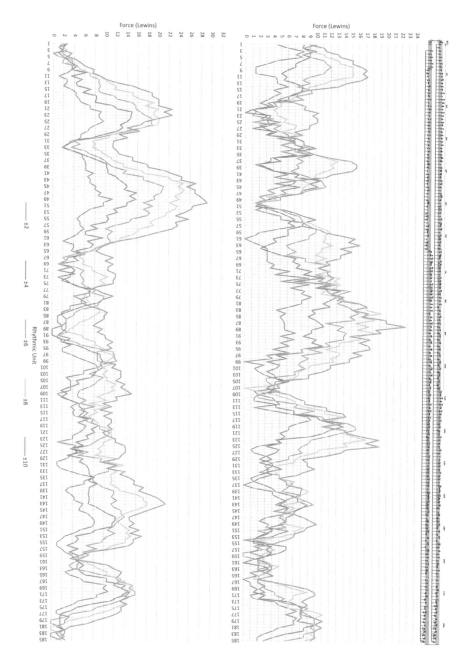

Fig. 5.11 Etude 18, RUs 1-185, force graphs. (**a**) (top): hexatonic force graph (force applied to Fourier Balance 3, measured in Lewins—span comparison). (**b**) (bottom): octatonic force graph (force applied to Fourier Balance 3, measured in Lewins—span comparison)

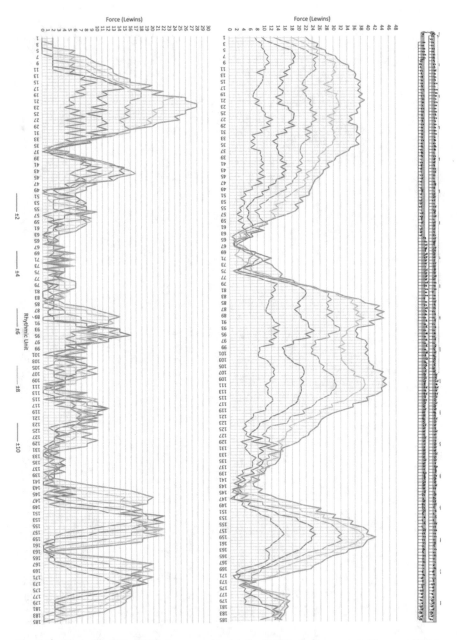

Fig. 5.12 Etude 18, RUs 1-185, force graphs. (**a**) (top): diatonic force graph (force applied to Fourier Balance 5, measured in Lewin—span comparison). (**b**) (bottom): whole-tone force graph (force applied to Fourier Balance 6, measured in Lewins—span comparison)

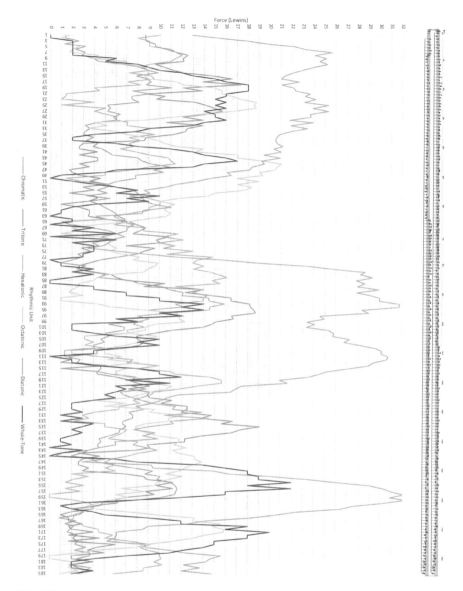

Fig. 5.13 Etude 18, RUs 1–185, force comparison (force applied to the six balances, using ±6 as a representative span)

voice moves to Bb rather than B, maintaining the purity of interval.[3] Right after the Bb we have our first interval other than a perfect fifth, a major second (as if the Bb causes a ripple effect, slightly disturbing the white-note, perfect fifth waters), and

---

[3] Unless otherwise noted, it is always the RH that is referred to when discussing a particular RU.

Fig. 5.14 Etude 18, RUs
186–196, piano roll (notes
in P-space, C3 = middle C)

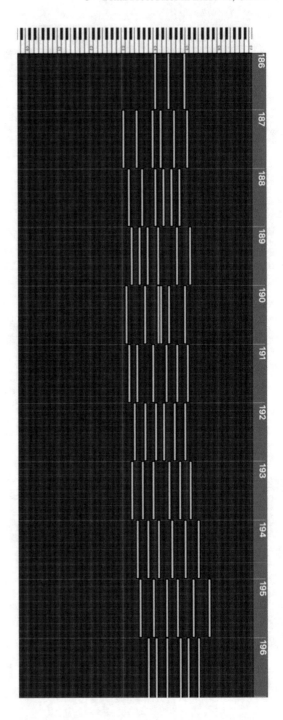

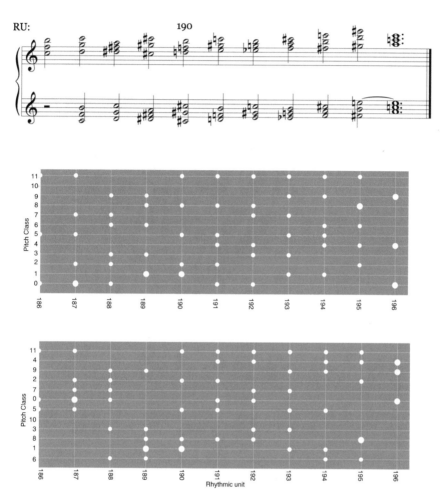

Fig. 5.15 Etude 18, RUs 186–196, comparison of two graph types (chromatic PC graph followed by Co5 PC graph)

even afterwards, all the way until the next non-white note (the Eb at RU 34) the majority of the ICs are 5s, with only a few 2s, 4s, the occasional octave or unison and a single 1 (the major seventh at RU 21) and 6 (at RU 24). This creates a very clean, open sound that is further enhanced by a voice-leading that often leaps rather than moving stepwise—especially at the very opening—serving to place these open, perfect fifths and fourths in relief.

The opening gesture is the only real upward motion in the *Vivace*—the rest moves slowly down the keyboard. Interestingly, the apex of these 18 RUs that move up an octave and a half are three RUs that are octaves or unisons, the last of which is an octave F, the highest point in the piece. That the registral peak of the work is marked by these most open of sonorities is not insignificant—as if the work is ultimately

Fig. 5.16 Etude 18, RUs 186–196, conical Co5 PC graph (notes placed in PC space ordered by the Co5 using polar coordinates: the y-axis is rendered as a circle, and the black rings mark the passage along the x axis)

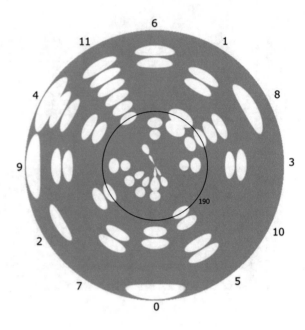

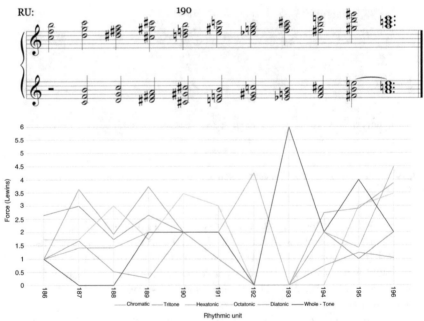

Fig. 5.17 Etude 18, RUs 186–196, force comparison (force applied to the six balances for ±0, i.e. without any spanning)

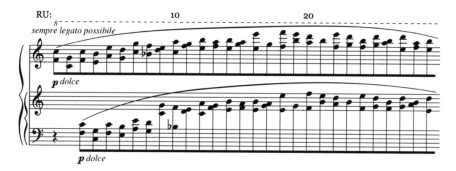

Fig. 5.18 Etude 18, RUs 1–29, score (repeat of Fig. 5.1). © 2005 Schott Music, Mainz—Germany

reaching for this purity and will have to again try to achieve it through the rest of the piece.

The importance of this moment is further highlighted by how the hands interact. The LH, lagging two eighth notes behind, at first doubles certain right-hand notes between RUs 13 and 17. When it then approaches its peak at RU 20, the RH complements it with two perfect fourths (at RUs 19–20) to form two minor triads. (It is also worth noting that the next triad is also a minor one, embedded in the seventh chord at RU 22.) The minor triad is the chord that will, after the winding journey of this brief etude, be finally achieved at the very end in the *Lento* conclusion, also at the end of an upward motion. The numerous ways in which that final chord encapsulates many features of the opening—diatonic implication, consonance, primacy of perfect fifths, even the reaching of a minor triad at the end of a climb—further attests to the idea that this is the home territory at which the work begins, departs from, occasionally alludes to over the course of the work and eventually returns to.

The clarity of the diatonic macroharmony in this opening is reflected in the dominance of the diatonic force relative to the others, as shown in Fig. 5.19. The Bb at RU 8 does not cause much of a weakening except for temporarily canceling out the effect of the E (as they are on opposite sides of the balance). Other sections in the piece will, however, show an even stronger diatonic force, due to a structural feature mentioned in the introduction: as the discussion below will demonstrate, the central sections of the palindrome are not based on full diatonic sets but rather on the 024579 subset that exerts an even greater force on Fourier Balance 5.

The purity of this opening scale area is also evidenced by the correlation strength in this stretch; the entire texture fits within the eight-PC set framework, the B-flat causes a slight disturbance with a set of seven, and even with a set of six there is a section of complete correlation (excluding the widest, ±10 span), given that, from RU 11 to RU 31, the passage omits C entirely, leaving a set of six PCs for this stretch (namely, the white notes other than C).

Omitting the C here in fact causes the second destabilization (the B-flat at RU 8 being the first), robbing the listener of the note that had functioned as the presumptive tonic. This presumption would have been based not only on the fact

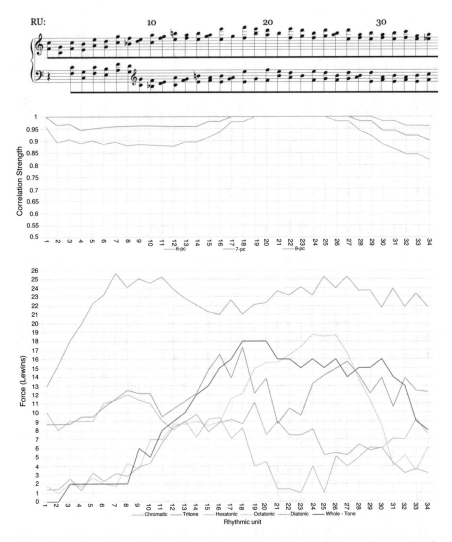

Fig. 5.19 Etude 18, RUs 1–34, comparison of two graph types (correlation strength comparison for ±6 and scalar force comparison for ±6)

that the PCs imply a C major scale, but also because C has been our lowest note so far, and this note is bolstered by a G above it when it appears, thus forming a kind of temporary bass note or anchor. This second destabilization sets the stage for greater chaos ahead. The sense of unease is heightened by the fact that the passage hovers above an E as a temporary bass note—unstable territory due to the minor-second E-F relationship the section is grounded on, building tension that calls for a resolution.

This is perfectly illustrated by the tritone force graph (Fig. 5.20), in particular the discrepancy between the forces of smaller and larger spans—which also serve to show the unique nature of the simultaneities at the registral peak in this passage.

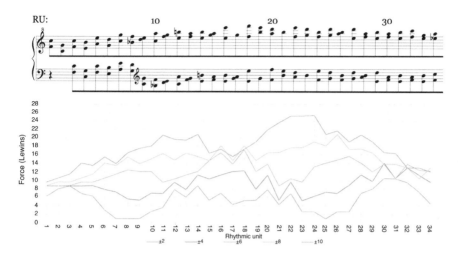

Fig. 5.20 Etude 18, RUs 1–34, tritone force (force applied to Fourier Balance 2, measured in Lewins—span comparison)

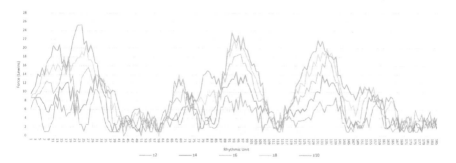

Fig. 5.21 Etude 18, RUs 1–185, tritone force (repeat of Fig. 5.10; force applied to Fourier Balance 2, measured in Lewins—span comparison)

While the ±6 span is reasonably (but not especially) strong here, the ±10 is at its most pronounced in the entire etude (see Fig. 5.21), peaking at RUs 22–24 where ±6 has its greatest dip. While the relatively even scalar distribution of PCs in the immediate vicinity of RU 23 don't register strongly on Fourier Balance 2, passages further out at RUs 11–15 and 28–34 (that get swept in under the larger spans) show a greater frequency of E, F, A and B, which with the addition of the Eb at 34 forms 3459E that exerts a strong tritone force; this is attested to by the peaks around these moments in the smaller spans.

These short sections of uneven distribution with strong tritone force highlight the instability of this section anchored on the E and F, proving a more sensitive barometer than the diatonic force graph of the inherent tension in this passage that seeks a resolution. Still more revealing is the fact that they frame a central passage that is different in constitution: none other than the registral apex that is harmonized

RU: 30   40   50

Fig. 5.22 Etude 18, RUs 30–59, score. © 2005 Schott Music, Mainz—Germany

by the minor triads. The fact that the more zoomed-in spans have stronger tritone forces before and after this moment—and plunge during it—shows the extent to which this moment is treated differently in the choice of PCs and hence to how this peak is placed in further relief.

The first modulation away from the diatonic macroharmony is an example of Ligeti's experimentation with various parameters that influence our perception of stability. As soon as the next non-white note (after the initial Bb) appears (the Eb at RU 34, triggering a steady increase in chromaticism) we get the long-awaited C at RU 35. This is no incidental passing note, instead appearing within a perfect fifth (as the C did when it first established its foundational function at the outset) and remaining a veritable anchor: it is the lowest note until RU 51 and appears no less than seven times in the interim (see Fig. 5.22). Hence, we get a sense of stability that to some extent mitigates the disorientation caused by our accelerating departure from the opening white-note set, creating a careful gradation in the departure to distant territory. This smoothness is enhanced by the fact that after accompanying the anchoring C at RU 35, G disappears entirely for a stretch of more than 20 RUs—the texture begins to shed the remaining vestiges of the previous stabilizers.

This chromatic departure is itself carefully calibrated. The Eb is the next PC down the Co5 after the Bb we have already heard (at RU 8). Next comes an F# at RU 36—the first new PC in the upward direction. A deliberate, stepwise widening of the gamut of PCs begins to reveal itself. After the F# comes an Ab at RU 44, the next PC downwards. The assignation of accidentals—PCs in the downward direction being flats, those upward sharps—further highlight Ligeti's use of the Co5 as the guiding principle in this motion.

As the range of PCs broadens in both direction, the pitch space register moves generally downward. As can be observed in Fig. 5.23, at first B, then E leave the texture as the downward area becomes more prominent—so much so that the F# could soon be considered a continuation of this descent (which is later on reinforced by the fact that it is actually respelled as a Gb at RU 65), a downward motion further extended by the later reentry of the B and E (at RUs 52 and 61, respectively), also now on the "other" side of the modulation.

However, the speed of the shift is not only gradual (relative to the considerably more rapid modulations later in the etude) but also lazy, in that several PCs of the

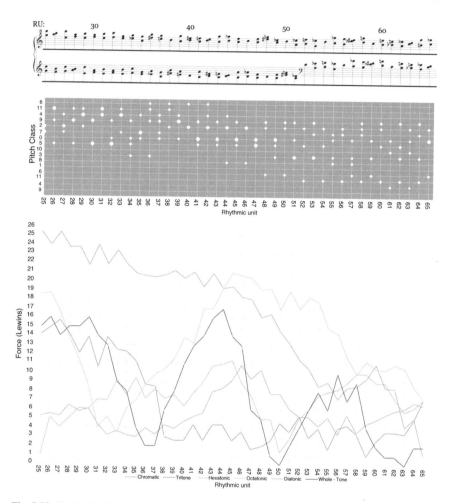

Fig. 5.23 Etude 18, RUs 25–65, comparison of two graph types (extended Co5 PC graph and force comparison for ±6)

opening macroharmony dawdle for a while, further adding to this sense of widening. Rather than occupying a total of seven steps (in the diatonic set), by RU 61 the texture has ballooned to 11, an almost complete cover around the Co5.

A few important changes accompany the macroharmonic shift. As soon as the texture becomes more chromatic, the nature of the intervals changes. First of all, they are much narrower—gone are the octaves and sevenths, now the widest interval is a major sixth. There is also a change in the relative frequencies of intervals, with many more appearances of tritones: in the opening white-note passage, there was only one (at RU 23), while after the pivotal Eb at RU 34—which itself enters on a tritone— many more appear, sometimes even on consecutive RUs.

In addition to the nature of the intervals, the type of motion and voice-leading also transforms considerably. Rather than the frequent leaps that characterized the opening, Ligeti now employs more restricted, stepwise motion, which in combination with the narrower intervals causes a shrinking of the registral range of the passage (see Fig. 5.22). This is an example of Ligeti's parameter exchanges: while the gamut around the Co5 widens, the textural range narrows. It is also evidence that passages of different scalar constitution are characterized by a number of other variables, such as certain types of intervals or voice-leading that, as later analysis will show more clearly, are in fact tied to the scale types that they come with—one does not appear without the other. (Chromatic saturation is, broadly speaking, also a scale type, even though it does not have a corresponding scalar force in the graph types.)

The increasingly disorienting, chromatic nature of this passage is clearly reflected in the force graphs (Fig. 5.23). Unsurprisingly, the diatonic force plunges. The progressive chromatic saturation (and hence more even scalar distribution) is evidenced by the relative weakness of the chromatic, tritone and hexatonic forces. Of particular interest, however, is the spike in octatonic force, which is the strongest force from RUs 45–59 and at its highest here in the whole etude.

While the listener may not interpret this passage as sounding octatonic, the graph actually serves as a perfect mirror of the unique combination of a widening *and* downward motion along the Co5 as well as some of the destabilizing processes at play. The increasingly frequent Eb and F#, absence of G for a considerable stretch, disappearance of E from RU 42 and reentry of the B in the extension of the descent at RU 52 all register in an 02 octatonic scale (i.e. [0, 2, 3, 5, 6, 8, 9, E]) that produces a very strong tilt on Fourier balance 4 peaking at RU 50.

The intensification of the chromatic process comes to a peak with a few important events at RUs 65–69 that serve as a swift turning point to subsequently move the texture back to diatonicity (Fig. 5.24). First of all, F# gets respelled as Gb as mentioned above, hence respelling of the so-far only sharp note as a flat and further highlighting the descent down the Co5. The next new PC (the last missing one until now) is 1 at RU 67, but rather than being spelled as a Db, it appears as a C#; simultaneously, the 3 above is spelled as D# (the first respelling of a flat PC), followed within the next two RUs by G# instead of Ab, A# instead of Bb and a return to the F# spelling for PC 6. All of this marks a clear turning point, signaling that the texture will soon start moving back up the Co5 to the white-note set.

Before the move back up, however, we have a final, clear move downwards; one can trace in Fig. 5.24 the almost perfectly ordered temporary disappearance of pitches down the Co5 starting with the A at RU 62, narrowing the gamut in a highly controlled fashion. If we for a moment continue the previous interpretation of the reappearance of the B and E (at RUs 51 and 62) after their hiatus as an extension of the downward motion, then in RUs 67–70 we have temporarily completely abandoned the upper home territory (F, C, G, D and A).

This shrinking of the gamut is soon accompanied by a similarly streamlined shift up the Co5. The G#—the furthest PC from the white-note set on the Co5—is the first to be abandoned after RU 68, while the last to leave are the nearest PCs, namely Bb at 73 and F# at 75 (maintaining in their last iterations their original sharp and flat

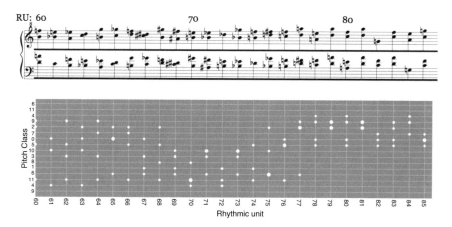

Fig. 5.24 Etude 18, RUs 60–85, extended Co5 PC graph (notes placed in PC space ordered by the Co5)

spellings, respectively)—as if reversing the harmonic widening motion of the first modulation in the etude. From RU 67 (the center of the peak of the chromatic process) until the end of the modulation only one white PC has a prominent role: B, a PC that extended the downward Co5 motion and now takes its place beside (or below) the black PCs. Two other white PCs appear once each, the E (as a passing note at RU 70) and F at RU 74, the boundary of the white-note set with the black notes below them on the Co5 and a herald for the snap back to the white-note set at RU 76, which happens with F and C (our home note and the next PC up the Co5).

Illustrating the density of the chromaticism of this passage is the fact that the correlation strength is at its weakest in the entire *Vivace* (Fig. 5.25a), reflecting not only the appearance of many PCs, but also that none of them prove particularly more salient and frequent than others: there are no anchor notes to give a semblance of key and we have reached almost total instability. The chromatic saturation is so thorough in this passage as to also cause the absence of any strong scalar force (Fig. 5.25b).

The listener feels this disorientation and lack of anchor not only due to the relative frequencies of the PCs, but also due to the nature of the voice-leading, an intensification of the shift that came about with the initial modulation from the opening white-note macroharmony. The more limited, stepwise motion now yields a dizzying texture of superimposed, mostly chromatic descending lines (highly reminiscent of Ligeti's Etude 9, *Vertige*—"Vertigo"—with its Shepard tone effect) (Fig. 5.26).

With the recapitulation of the white note set at RU 76 two important, rapid shifts take place. The first is a sudden return to the type of intervals that defined the opening—for a string of 13 RUs we have nothing but perfect fourths and fifths (namely IC 5), save for one unison and one major second—as well as the sharp-edged, leaping voice-leading that now stands in stark contrast to the vertiginous sliminess of the descending chromatic lines in the immediately preceding section.

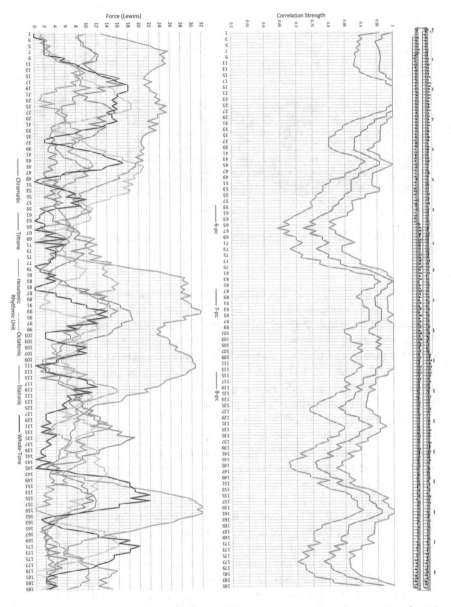

Fig. 5.25 Etude 18, RUs 1-185, comparisons. (**a**) (top): correlation strength comparison for ±6 (comparison across 6-, 7- and 8-PC sets using ±6). (**b**) (bottom): force comparison (force applied to the six balances, using ±6 as a representative span)

Fig. 5.26 Etude 18, RUs 60–88, annotated score (color highlights for descending lines). © 2005 Schott Music, Mainz—Germany (unannotated)

The return to the white-note set has brought with it the specific texture that the listener now associates with this macroharmony.

The second is the sudden surge in diatonic force (to be addressed in greater detail below and illustrated by Fig. 5.28). While this is to be expected, the magnitude of the rise is unique, surpassing that of the opening. This is due to the aforementioned shrinking of the gamut around the Co5: as soon as we hit the white note set at RU 76, B is nowhere to be seen (and will not appear for another 44 RUs, next appearing at RU 120). The reason for this becomes clearer if we keep in mind that this PC was the first extension of the downward motion beyond F#, as well as the most frequently-appearing white note in the passage from the chromatic peak up to this recapitulation (namely the portion at RUs 67–75): it ended up taking a function removed from the white note set, and instead became grouped with the foreign black notes. Thus Ligeti has set up an exchange between 024579 and 1368TE, two opposite sides of Fourier Balance 5. At the recapitulation in RU 76, the absence of the B in the white note set yields the PC set C, D, E, F, G, and A that exerts perfect pressure on the balance without any pitches that cancel out the force (given the absence of any tritone relationships, which are on opposite panhandles), resulting in the remarkably strong showing for the diatonic force—which is in fact about to grow even further due to the unique nature of the ensuing transition.

It is worth noting two important points about the 1368TE area. First, it is somewhat obscured as it is embedded within the two rapidly alternating Co5 processes of narrowing and upward motion, and hence does not have the time to truly establish itself as a clear, separate macroharmony (and therefore does not appear strongly on the diatonic force graph shown in Fig. 5.19, though a not insignificant rise appears in the smaller spans around RU 71); nevertheless, it is a clear and direct product of both of these modulatory processes. Secondly, though we did also consider PC 4 to have moved from its original position in the white-note set to an extension of the downward move, it arguably does not turn the implied 1368TE hexachord into a diatonic scale: it only appears once in this modulation back to the white note set, and clashes with the appearances of PC 5 (which to the listener, could be interpreted as E#) that it is closely sandwiched by, creating ambiguity as to a possible seventh degree.

With the harmonic narrowing process we've entered the central section of the palindrome presented in Fig. 5.3. The modulation back up to the white note set was considerably more compact in its Co5 span than the first modulation away; the same applies to these new macroharmonies, occupying fewer steps around the Co5 than the opening, full diatonic set. It is as if the moving "ribbons" or bands that track the upper and lower borders of the motion around the Co5 have moved closer together, for both the transitions and arrivals.

The recapitulation of the white-note set has not only brought about a return of the opening scale area, voice-leading, texture and so forth, but also some of the initial modulatory processes: the first foreign notes are again a Bb (at RU 87) followed by an Eb (RU 91—see Fig. 5.27). It's soon clear that the passage is again modulating; it seems as if the etude is going down the same path as it did once already. However, the texture actually stands at a crossroads. Now that the two first flat notes down the Co5 have been reintroduced, an option is to continue the chromatic widening in both directions as before—or to keep moving down the flat direction only, which is what happens now. Not only are there no sharp notes introduced, but, as the passage moves in the flat direction, pitches of the white note set begin to disappear one by one: the modulation continues the pattern of narrowness (Fig. 5.27). As Db, Gb and Ab join the texture, PCs E, A and D leave entirely. G and C become more infrequent, getting cut off RU 111 and leaving the focus of activity on the increasingly dominant PCs 1, 3, 5, 6, 8 and 10: another manifestation of the six-PC diatonic subset.

Fig. 5.27 Etude 18, RUs 89–138, score. © 2005 Schott Music, Mainz—Germany

For the second departure from the white-note set, we see that Ligeti has chosen a type of motion that contrasts from the first not just in procedure but also in destination. The streamlined nature of the motion—occupying a narrower width on the Co5—yields a similarly clean arrival on an easily identifiable scale area (RU 112). Similarities with features of the opening also appear in the voice leading: there is a clear preference for leaps. In fact, this idea is actually exaggerated to such an extent that we end up with an unusual middle ground: the jumps are so pronounced as to cause the two voices to seemingly split into four, creating a kind of rhythmic hocketing through RU 106.

This Escher-like aural illusion has two interesting effects. The first is that each half of this texture (namely the upper and lower pairs of voices) ends up moving in a largely stepwise manner. The second is that this type of alternating motion yields a combination of both the open IC 5s and narrower, less consonant major and minor seconds and thirds. Hence, this transitional texture that is a combination of previous passage types (in that it is both narrow in span around the Co5 yet moves in a downward direction) also occupies a kind of middle ground when it comes to interval and voice-leading, or, to be more precise, manages to retain features of both simultaneously.

The intentionality of this modulation and arrival at the next macroharmony is highlighted by the first dynamic contrast of the etude. As soon as the last foreign notes, G and C, finally leave the texture for a while at RU 111 and allow the new scale area (here with PCs 13568T) to be perceived by the listener unhindered by other PCs, we get two accents, fortissimo, marking an emphatic establishing of the new macroharmony (RUs 112 and 114 in Fig. 5.27). These accented dyads establish a framework for those further on in the etude: they are almost all IC 2s (two important exceptions are both major sevenths) and move downwards, reflective of the motion of the rest of the texture in the *Vivace*.

Further marking the clean arrival of the reduced diatonic scale area is a brief return to the same type of voice leading that has so far been associated with the white-note set, establishing a connection between this type of texture with any clear iteration of the diatonic scale (or its six-PC subsets)—not just in its original, white-note key.

As shown by Fig. 5.28, the narrowing of the span around the Co5 throughout this part of the etude—from the recapitulation at RU 76 and the ensuing modulation through the firm arrival at 13568T during RUs 112–118—is evidenced both by the high correlation strength and a surge in diatonic force that is in fact at its highest point in the piece. (For comparisons across the etude, see Fig. 5.25).

The most significant dip in the force in this passage happens when G yields to Gb at RU 99—a revealing moment not only for the motion of this passage, but also a herald for how the texture will move to the next part of the work. The Gb/G exchange is an important factor in this stretch, as the G persists for quite a long time after its upper Co5 neighbors D, A, E and B have disappeared (see Fig. 5.28)—and only stops once the Gb kicks in at RU 99 and supplants it, initiating the main motion within this modulation. The G natural at RU 110 is the only cancellation of an accidental in this entire passage, and the next foreign note is again a G at

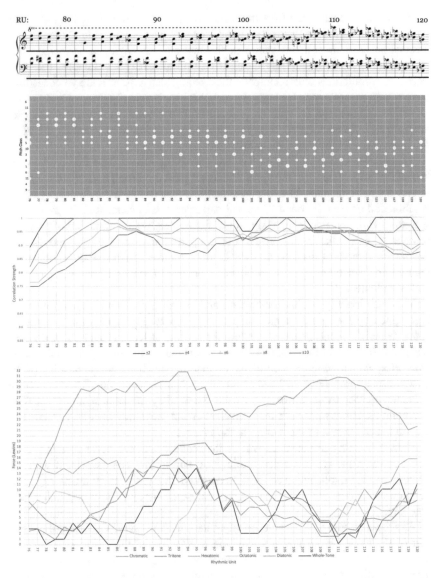

Fig. 5.28 Etude 18, RUs 76–120, comparison of three graph types (extended Co5 PC, 7-PC correlation strength and force comparison graph for ±6)

119 where it marks the beginning of the next modulation, the final "zag" back up the Co5 in the *Vivace*. The pivotal role of the G may also be a reason why the Ab was strangely late in appearing when the modulation began at RU 87, bucking the otherwise perfectly ordered trend of new PC appearances down the Co5: the G, without any semitone-neighbor PCs, was placed in greater relief, further

accentuating the moment when is replaced by the Gb (after which the Ab promptly makes an appearance).

The entry of the G natural at RU 119 provokes a modulation back up the Co5 via a harmonic widening. The G is followed by a B at RU 120 (one Co5 step down from the current 13568T-area), a D at 122 (next step up) and E at RU 125 (next step down)—a neat widening around the Co5, shown in Fig. 5.30. This resembles the processes of the first modulation away from the opening macroharmony (starting at RU 35), as does the voice-leading: we are back to the vertiginous, descending chromatic lines.

Ligeti sets up another structural delineation to mark our return to the slimy textures. The G and B at RUs 119–120 both appear as parts of tritones, the first occurrences of this interval since RU 74—which was during the last such passage. Not only does this moment create a stark clash with the string of IC 5s just before it, but also signals a return to these kinds of textures. This is further illustrated by two types of graphs in particular. The first is the correlation strength which takes an enormous dip in this moment to levels only reached in that first modulation; the plunge is especially deep in light of the recently shallower trenches during the central modulations between the diatonic hexachord-based sets. Secondly, the diatonic force also displays the same steady weakening that was apparent in the first modulation away from the opening macroharmony (compare the similarities between RUs 30–65 and 110–145 on Figs. 5.12 and 5.25). Hence, after the central, "narrow" hexachord-based passages in the *Vivace*, we have moved one step outwards on the palindrome, the two "wider" outer modulations forming a clear symmetrical pair— one moving down, the other up.[4]

As mentioned above, there are two cases where accented dyads form major (rather than minor) sevenths. Unlike their IC 2 counterparts, in both situations these more dissonant versions appear in destabilizing contexts. The first is the lone accent at RU 137. Not only is it embedded within a densely chromatic, transitional passage, but also prepares an important textural shift marked by the first sharps (since RU 75) at RU 139. Ligeti again uses accidental assignations to signal harmonic movement. Exactly in the manner observed at RU 67, these sharps mark the precise moment the texture begins its direct ascent up the Co5. As with the mirror passage, the modulation is more sluggish, maintaining a wide span around the Co5 rather than the streamlined, narrow movements that linked the central hexachord-based stretches. The wideness of the span however makes the modulation no less methodical: the F# and G# at RU 139 are followed immediately by a neat, stepwise shift up the Co5 with a D and A at RU 140, E at RU 144 and finally B at 150, the last two now in their original position in the upper portion of the white-note set on the Co5 (Fig. 5.30).

In the manner they were used at RU 112, a series of accented IC 2 dyads mark our arrival at a new macroharmony at RU 150, in this case the return to the white-note set (see Fig. 5.29). Two important events take place here. First of all, the first accented

---

[4] See Fig. 5.3 for reference.

Fig. 5.29 Etude 18, RUs 113–159, score. © 2005 Schott Music, Mainz—Germany

PCs include a B: its emphatic inclusion in the new scale area means we have returned to the full white-note set of the opening, rather than the more restricted 024579, reinforcing the palindromic nature of the structure in the *Vivace*. The second is the change in texture. Given that the accented IC 2s are now appearing as major seconds rather than minor sevenths, we have a return of the hocket texture. Observing the PCs at play reveals this is no accident: the Eb is late in clearing out of the texture and is still present between the first two accents. Hence this passage is something of a bridge, straddling a white-note set with another key area—exactly the kind of moment this type of texture first appeared.

An interesting visual illustration of this is the surge in whole-tone force during this moment, as shown by Fig. 5.30: the ten pitches encompassed by the accented outburst form 3579E, all belonging to whole-tone set 1. Responding strongly to the late exit of Eb, this metric makes even more evident the partly transitional function of this passage.

This interesting leak of the Eb is part of a larger pattern. As the Fig. 5.29 shows, Eb was the last foreign note before the accents (and maintained its flat spelling despite the sharpening of most other black PCs). It was also the PC to kick-start the very first modulation in the etude—and shall also prove to play a key role in the *Vivace*'s final modulation. (While the Bb at RU 8 was the first note in the piece that did not belong to the opening macroharmony, its momentary appearance did not

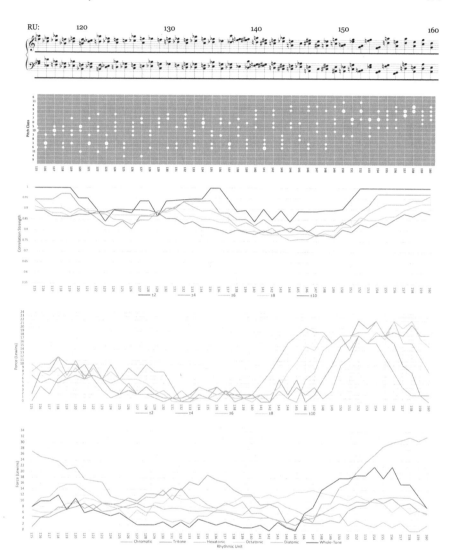

Fig. 5.30 Etude 18, RUs 115–160, comparison of four graph types (extended Co5 PC, 7-PC correlation strength, whole-tone force and force comparison for ±6)

cause an immediate change in harmonic course—that would only happen with the Eb some 25 RUs later.)

Following this, the return to the voice-leading of the opening, now with a string of six perfect fifths, marks our last moment of stability in the *Vivace*—before a tritone again provokes a modulation. It is worth noting that while there is a surge in diatonic force in this moment of stability (Fig. 5.30) it on the whole weaker than in the central hexachord-based passages, especially in the higher spans (see Fig. 5.25 for the comparison). This is not only due to the relative brevity of this episode but also the greater distance around the Co5 (caused by the accented B as well as the intruding Eb); this visual analysis further highlights how this passage closely mirrors the opening in constitution (rather than the central reduced-diatonic passages), closing the palindrome and setting the stage for the third zigzag that will function as a conclusion.

Though this last modulation in the *Vivace* displays several features that are similar to previous ones in the etude, it differs in a few crucial aspects. Most obviously, it does not segue smoothly into another passage, but rather seems to hit a brick wall (Fig. 5.31). The theatricality of this moment is heightened by the fact that Ligeti instructs the pianist to then repeat the entire *Vivace*, now *Prestissimo*: failing to find a way forward, we restart the machine and drive forward even faster in the hopes of finally breaking through. (In a sense, we do—on the other side of that wall is a different sphere altogether, the *Lento*.) This passage is the only movement down the Co5 that is dominated by sharp (rather than flat) PCs, and the only modulation away from the white-note set to arrive at another clear, full diatonic set, rather than a narrowed or obscured one. Finally, there is an enormous *crescendo* in the passage, leading to *fffff* in the final crash—which is particularly significant as it is a dynamic level greater than any of the accented dyads. All these features make this last part of the *Vivace* something of an outlier, separate from the rest of the sections that constitute the central symmetrical structure of the work.

As mentioned above, the modulation is initiated again by a tritone—with none other than a Bb, once more the first foreign note, the first step down the Co5. Importantly, the Eb we had heard only a couple of RUs ago as the intruding PC in the white-note set is skipped, the next new PC being an Ab at RU 168. It is after this point that the rest of the accidentals are marked as sharps, not flats—with one important exception.

Accented intervals resurface at RU 172—now *fortississimo* in the surrounding *mezzoforte* texture—but the second dyad is a major (hence, destabilizing) seventh rather than a minor one, presenting a further sign that this modulation is somehow veering off course and our progress may be doomed. While the accented minor sevenths always belonged to the macroharmony the passage in question was

Fig. 5.31 Etude 18, RUs 160–185, score. © 2005 Schott Music, Mainz—Germany

establishing, the sharpened A at RU 174 (turning this seventh into a major one) is a "wrong" note that clashes with the E-major scale this section will eventually arrive at (which namely has an A natural).

To extend the driving metaphor, this accented A# actually causes a swerve in the hurtling down the Co5, leading to a patch of four RUs that negate some of the progress. This moment includes appearances of a D, G and C, which, as shown by the extended Co5 PC graph in Fig. 5.32, are pitches that violate the otherwise remarkably orderly exit of PCs down the Co5. (It is for this reason that this

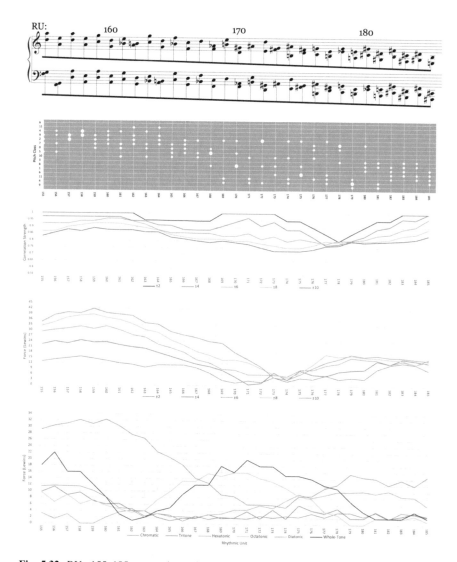

Fig. 5.32 RUs 155–185, comparison of four graph types (extended Co5 PC, 7-PC correlation strength, diatonic force and force comparison for ±6)

modulation is labeled as "compound" in the structural overview in Fig. 5.3; while the motion itself is narrow, the persistent presence of the foreign PCs muddies the waters, retaining an element of the wider motion type.) In fact, this moment also includes an actual flat PC: none other than the intruding Eb, which disappeared since its last intrusion in the white-note set back at RU 151 and was conspicuously absent between Bb and Ab at the start of this final downward modulation. This Eb seems to repeatedly function as a kind of thorn in the side of new scale areas—and, as shall be discussed below, has yet to complete its function in this regard.

The passage gets back on course immediately after the Eb, the next RU sharpening it to an E natural, anchored on an A natural a fifth below. The A, meanwhile, corrects the A# from the last accented dyad. As shown by Fig. 5.32, by this moment PCs 0, 2, 5, 7 and 10 have one by one left the texture while 4, 11 and finally 9 have assumed their position as extensions of the modulation down the Co5: we at last reach our final macroharmony of the *Vivace*, a clear E-major scale. This moment is demarcated with the immediate reentry of the opening texture, a string of six perfect fifths beginning at RU 179, with leaping voice-leading. It is also here that Ligeti marks an extra hairpin to further accelerate the crescendo, making the dynamics of this macroharmony balloon to *fffff*, the loudest in the etude.

This sense of gaining of momentum in this last modulation is evidenced by the correlation strength (Fig. 5.32), which when zoomed out (at ±10) is at its weakest in the whole etude, reflecting how rapidly the modulation is taking place; there are no stops along the way where certain PCs establish some prominence. It is instead a mad dash down the steep slope of both the Co5 as well the piano's registers: as shown by the piano roll (Fig. 5.33), this passage is the fastest movement down the keyboard in the etude, a fusion of harmonic and registral motion that will prove to be important to our understanding of the conclusion of the work.

The sheer speed of harmonic motion in this last hurtle is further displayed by the weakness of larger spans of scalar forces across the board, shown in Fig. 5.32. While this is partly due to the empty space on the other side of the section boundary (that is, when using spanned metrics, the sizes of the sets analyzed grow progressively

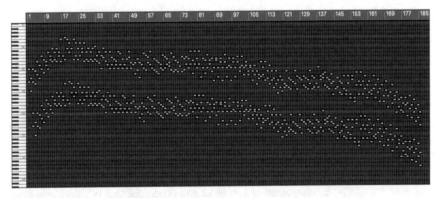

Fig. 5.33 Etude 18, RUs 1–185, piano roll (repeat of Fig. 5.4; notes in P-space, C3 = middle C)

smaller as we reach the last datapoint) this alone does not account for the extent of the weakening of scalar forces in this final passage. On the other hand, smaller sets still provide valuable data, such as the spike in diatonic force at the end, with the ± 2 span being at one of its strongest levels in the etude: a clear indicator of the establishing of the final macroharmony in the *Vivace*. It is worth noting that the whole-tone force also provides a valuable glimpse into the unique nature of the modulation, picking up on the disturbing D natural (late to exit and involved in the post-accent swerve) that causes a spike in the force at RU 172 from the implication of whole-tone scale 0 (largely due to the presence of D, E, F#, G# and A#).

The gathering in this modulation of immense momentum—in registral and harmonic motion as well as volume—leads to an increase in actual velocity in the form of the *Prestissimo* repeat, making the stasis of the *Lento* that follows even more startling and powerful to the listener. At this moment the listener may think that we have suddenly entered a new realm and have left all of the goals and procedures of the *Vivace* behind. However, on closer inspection it turns out that the *Lento* is in fact one last attempt at achieving the etude's aim of landing—permanently—at home ground.

As discussed at the outset of this chapter, the white-note set was clearly established as the home territory at the beginning of the etude. Despite all its attempts, the *Vivace* failed to return to this macroharmony in a way that it could finally come to rest there: it always ended up being destabilized and sliding down the Co5 again. The final modulation did indeed reach a full, diatonic set at the end, the only downward spiral to do so—but not the right one, with the final modulation arriving at an E-major scale. We've ended up on the wrong side of the looking glass.

The etude so far has been in large part a drama of direction. It begins with an optimistic leap upwards—but then gravity kicks in, pulling the rest of the *Vivace* inexorably down to the abyss. The first modulation away from the white note set goes down the Co5, and despite every push back up the returns to the home territory are always short-lived, pulled repeatedly back down into foreign key areas. It is as if, after many failures, the final modulation in the *Vivace* tries something different. Despite moving once again down the Co5, for the first time it largely employs sharps instead of flats in the hope that this may yield different results through the attempt to move in the direction of a sharp key area—perhaps a finally successful and perma-nent move up. It instead results in catastrophe, causing an acceleration of the downward motion and the eventual crash, prompting another, considerably more breathless attempt at the whole venture. However, the careening *Prestissimo* is still doomed to its fate, failing to escape the force of gravity. (Ligeti's performance markings are telling: for the *Vivace* he marks "fluctuations in tempo ad lib.," but for the *Prestissimo* he writes "if feasible without hesitation.")

This idea of being trapped and unable to escape a certain fate was a recurring motif in Ligeti's oeuvre, one that was in large part informed by his own experiences through horrors of twentieth century history. Some of the most notable manifesta-tions of this idea are in some of his earlier piano etudes, with insistent, often crazed hammering of the extremes of the instrument's register, as if in an attempt to break out from the boundaries of a metaphorical prison. (See, for example, his Etude

13, *L'escalier du diable*—"The Devil's Stairway.") The ending of the *Vivace* in Etude 18, despite not being at the very lowest notes of the keyboard's lower register, is another example of this: at the end of this inexorable descent there seems to be an invisible wall that cannot be breached.

Yet in this very last work that Ligeti wrote, we do at last manage to escape: there is an astonishing moment of transfiguration. While it may be fanciful to speculate on any extra-musical meanings of this moment, it is tempting to read into it a kind of epigraph from a different state or realm of being (a presumably posthumous one). Ligeti did not adhere to any religious belief system that would render this idea at all more valid (Sabbe 1979, 16), but the nature of the *Lento* does have an element of the "celestial chorale" that makes it reminiscent of moments such as the "wave from afar" concluding Schumann's cryptic penultimate work, the *Gesänge der Frühe* ("Songs of Dawn") Op. 133.

The *Lento* finally achieves what the *Vivace* failed to do: arrive home. In order to do so, it employs a different procedure. As mentioned in the opening to this chapter, it is the only modulation not to use the Co5; Fig. 5.34 shows no discernible Co5 pattern that could summarize the motion in the entire section. Given that motion up the Co5 never resulted in a permanent arrival, the *Lento* instead moves upward in *register*, finally countering the downward spiral of the *Vivace*.

Despite this change in procedure, there are many important parallels with the previous modulations. There is of course the reaching of the diatonic, white-note home area—here represented by the unison, A minor triad—at the end of an ascent. Then there is the use of sharps instead of flats for a passage that functions as an "upward" modulation. This feature in a sense makes the *Lento* start from where the

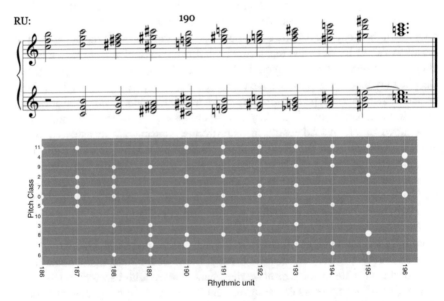

Fig. 5.34 Etude 18, RUs 186–196, Co5 PC graph (notes placed in PC space ordered by the Co5)

Vivace left off, given that the final downward motion was also dominated by sharps. Crucially, they share the one intruding flat PC: the Eb. As can be seen in Fig. 5.34, the only non-sharp accidental is the Eb at 192; this moment in the *Lento* is also significant for another reason, marking the only moment where all three voices move downward. In the rest of the passage, at least one of the voices move up from one chord to the next. The number of voices that move up from one chord to the next are most often three in number (to RUs 187, 191, 193, 194, 195)—resulting in the generally smooth upward trend—followed by one voice (to RUs 188, 190 and 196), two (to RU 189) and finally zero (to RU 192, the moment with the Eb). Thus the Eb is used to signal a move in the wrong direction, much like it did at the end of the *Vivace*—though here it is a registral wrong turn, whereas in the *Vivace* it was one on the Co5.

It is for these reasons that, despite the change in register, dynamics and texture, the final descent in the *Vivace* (RUs 159–185) and the *Lento* can be considered something of a pair, the third and final zigzag that, due to the unique features of both the descent and ascent that distinguish them from the previous modulations, stand apart from the main structural palindrome of the work not only in positioning but also in constitution.

However, the unique nature of the final return to the opening macroharmony creates a larger, more fundamental symmetry encompassing the entire etude. The *Lento* ends not with just any representative of the white-note set but a particularly distinguished one: an A-minor triad in root position. The fact that this is a unison chord is significant. First of all, the texture has to violate the canon, the LH having to skip a step to join the RH on this final chord. More importantly, this unison root-position triad refers back to the first two-handed chord in the entire etude: a unison perfect fifth.

Not only is Ligeti's *Canon* bookended by diatonic, white-note consonance framed by unison perfect fifths—as mentioned in the discussion of the opening, the fact that this is a minor chord is also significant. The registral apex of the etude, the peak of the only other successful motion up the keyboard in the work, is also marked by minor triads at RUs 19–20, a crucial aural landmark for the listener in the opening macroharmony; the eventual arrival at this very chord type at the end of the etude heightens the sense of homecoming.

Interestingly enough, the beginning of the *Lento* itself opens with the white-note set, the first two RUs covering PCs 0257E. These two chords also relate back to the very opening of the etude, as well as the conclusion of the *Vivace*, in that they are based primarily on IC 5: they are both quartal harmonies, stacking mostly perfect fourths (plus a tritone). They also create another mini zigzag within the *Lento* that starts at the white-note set, moves away and then back—a fractal-like local iteration of larger patterns that was an obsession of Ligeti's in his late work. A further layer to this pattern is the fact that the two times both hands are playing quartal chords simultaneously in the *Lento* are the second RU (i.e. RU 187, the first time the two hands play together in this passage, given the canon) and the penultimate one (the final chord leading to the A minor triad) which is also a combination of perfect and augmented fourths—another symmetry that frames the passage.

This pattern is reflected by the force graph in Fig. 5.35, which looks at the variance of the scalar forces from one chord to the next. As it uses ±0, there is no smoothening out in the data caused by broader spans, so certain forces display steep rises and plunges depending on the forces exerted by each chord, such as the whole-tone force that spikes due to whole-tone 1 set at RU 193, or the hexatonic force from the 03 hexachord at RU 192. There is however one force that has an unusually smooth curve, charting a steadier path over the course of this passage: the diatonic. It starts out as the generally most present force in the first five RUs, becoming gradually weaker from RU 189 until completely disappearing at RU 192–193 (the two hexachords mentioned above cause a tilt of zero on Fourier Balance 5) and then growing, step by step, until the final chord: a small zigzag and perfect illustration of the fractal-like phenomenon.

Though harmonic motion between individual chords is not the analytic topic of this book, it is worth briefly discussing the curiously tonal relationship of the penultimate chord to the final A minor triad. Looking at both hands together, the inner four voices form a standard $V^{4/3}$-I progression with conventional voice-leading: the second degree (B) moves to the third (C), the fifth (E) is held, the sharpened seventh (G#) moves up to the tonic (A) and the fourth degree (D) moves down a step to the third (C). The outer voices are a little less neat, with the low F# being part of a previous upward motion towards G# that gets smudged into the penultimate RU due to the canon delay, while the upper G#'s leap down to E is not

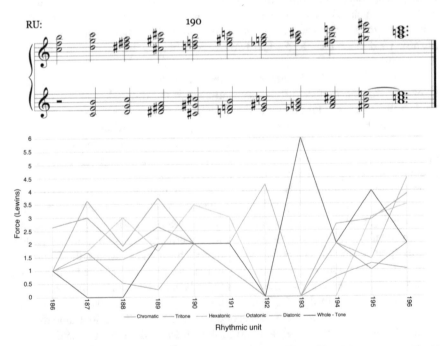

Fig. 5.35 Etude 18, RUs 186–196, force comparison (repeat of Fig. 5.17; force applied to the six balances for ±0, i.e. without any spanning)

standard voice-leading in the conservative sense of the term (though the pitch itself still belongs to the core V^7 chord). Nevertheless, this is a clear perfect cadence, a dominant-seventh to tonic progression—another fractal iteration, at the smallest level, of the forces of dissonance versus consonance, tension against resolution that guide the entire work, and a further indication of this etude's underlying "tonality."

This goes against the general understanding of Ligeti's use of triads in his late work as simply "surface sonorities" without any harmonic function, the composer having imported merely "the vocabulary but not the syntax of tonal music" (Searby 2001, 216). Instead, it furthers Shaffer's conclusions regarding Ligeti's use of triadic harmony discussed in the introductory chapter, namely that

> Ligeti composed meaningful harmonic successions, that those successions can be said to be syntactic, that the structure of those successions and the properties of those syntaxes have a strong relationship with some fundamental aspects of the successions and syntax of common-practice tonal music... and that understanding that relationship is fundamental to understanding the harmonic and formal structures of these works (Shaffer 2011, 193).

It would be difficult to here make the claim—as has been made for a lot of Ligeti's "triadic" writing—that the final two chords merely *resemble* a perfect cadence given the way our ears have been molded by tonal music, and that they are otherwise simply simultaneities comprised of pitches chosen for their sonic effect. Not only is this progression almost explicitly functional, but within the framework of an etude driven by tonal procedures their cadential, resolving purpose is without doubt a clear continuation and conclusion of the dramatic and harmonic narrative of the piece. As the landing point of Ligeti's final work, this A-minor conclusion is a truly poignant statement for a composer unique in the nature and breadth of his stylistic transformations—and a veritable testament to the remarkable direction these very last works of his oeuvre took.

Chapter 6
Conclusion

Beyond the significance of the A minor triad in *Canon* as the final chord of the piece (or indeed of the last work in Ligeti's oeuvre) is its relation to the previous etudes of Book 3. As has been intimated in earlier chapters, there is something of a meta-harmonic progression taking place over the course of these four etudes. The conclusion of *White on White* is anchored on the A minor triad; *Pour Irina* ends with a section that articulates a movement from an A minor triad to an E major one; in *À bout de souffle*, the E minor triad (the first rhythmic and acoustic consonance of the work) announces the conclusion of the etude; and finally, *Canon* comes to a halt on an A minor triad. In each of the etudes, these triads are outliers in the texture, coming across as moments of particular importance to the listener. They also all take place in the conclusions of each etude, essentially functioning as capstones or final destinations.

With the movement from A to E and back to A, these acoustic beacons seem to form something of a I-V-I progression over the entire course of the final book of etudes. Interestingly, the nature of the harmonic structure and dramatic arc of each of the etudes seem to reflect their position in this broader progression. There is something open-ended about *White on White* and *Pour Irina*, with the movements away from their opening macroharmonies—reflecting the initial movement away from home territory, namely I-V. In fact, the two seem to occupy different stages of this opening-up, the gentler fuzzification of Etude 15 announcing a departure from I and the wholesale, pronounced movement in Etude 16 marking a clear arrival at V. On the other hand, *À bout de souffle* and *Canon* have structures that circle back to the opening scale areas, with Etude 17 taking steps toward closure (but failing to fully do so) and Etude 18 finally, irrevocably arriving home.

Despite the highly conspicuous nature of these tight harmonic relationships across the etudes, their deliberate calibration on the part of the composer is merely hypothetical; we have no indication that the four etudes were conceived as a set, and the 6 years between the composition of *White on White* and *Canon* would seem to imply otherwise. What is clear however is the extent to which these final works not only move away from chromatic saturation and primarily reside in diatonic (and closely related) scale areas, but that this is accompanied by a manner of juxtaposition of macroharmony that establishes functional relationships between the scale areas, articulating alternations between stability and instability that define the dramatic narratives of each etude.

Perhaps the most interesting manifestation of this is the significance of fifth-relationships and Co5 motion in general throughout the etudes of Book 3. The fifth is ubiquitous, whether in the manner in which macroharmonic correlation changes, in the modulations between scale areas or more generally in the broader relationship of the primary macroharmonies in the etudes. These all reflect the nature of the majority of the scale areas themselves, given that the diatonic scale is itself a product of fifth-relationships. Ligeti has not only restricted the scalar gamut but has also derived harmonic procedures inherent in the scale type used.

This gives rise to the question of whether use of the diatonic scale has automatically brought with this a host of other functions as part of the package, somewhat in the manner that Schoenberg described for the use of the triad—"to introduce even a single tonal triad would lead to consequences. . . [it] makes claims on what follows, and, retrospectively, on all that has gone before" (Schoenberg 1926, 263)—or whether these diatonic features can function independently of historical and/or tonal implications.

To answer this, one can to a certain extent transplant the debate on Ligeti's use of triads—namely, whether they are merely coloristic in function or in fact have a syntactic role—to his use of diatonicity (and treatment of scale area in general) in order to address how we can reconcile the seemingly retro nature of the procedures in Book 3 with Ligeti's own statements. It is important to keep in mind that the description of Ligeti's late style as one that uses certain features of music of the past (such as triads or diatonic scales) *without* much of the accompanying baggage—advanced by scholars such as Drott (2003), Searby (2001) and Steinitz (1996, 2003)—is one that largely derives from Ligeti's occasionally somewhat misleading pronouncements. As mentioned in the introduction, this issue is perhaps best addressed by Shaffer, who discusses how Ligeti's "legacy-building" prompted a framing of the discussion around his late compositions that was not always a wholly accurate reflection of the musical style itself, given that "Ligeti desires to be seen as a 'late' composer—both in terms of his own career, and in terms of the broader history of music" which "would emphasize. . . [a] break from the now-exhausted material of the past" (Shaffer 2011, 209).

What is particularly interesting about the related question of Ligeti's use of scale area is his evolving relationship to it even within this late style. Shortly after composing Etude 15 Ligeti described his music as "diatonic. . . and not yet tonal" (Ligeti 1996, 11); while one can debate the extent to which this is an accurate characterization (per Shaffer above), it would be largely fair to say that Ligeti's etudes up to that point are at least not quite tonal, even if they exhibit a greater use of tonal syntax than is generally recognized.

However, imagining a spectrum between atonality and tonality on which Ligeti's music lies, the final etudes display a clear shift towards the latter, both within Book 3 and relative to the earlier works of his late style. While *White on White* indeed

Fig. 6.1 Etude 7 (*Galamb Borong*), bars 49–56, score. © 2005 Schott Music, Mainz—Germany

manages to largely avoid tonal implication despite its strict adherence to diatonicity, the opening of *Pour Irina* written 2 years later signals that Ligeti is no longer making a case of evading tonality; *À bout de souffle* and *Canon* go on to articulate structures derived from the presence of clearly-defined, harmonically-functioning tonic key areas serving as the home ground that the textures depart from and ultimately return to.

Of Ligeti's earlier etudes, five can be said to use limited, strictly-defined scale areas extensively: Etudes 1 (*Désordre*), 7 (*Galamb Borong*), 10 (*Der Zauberlehrling*, "The Sorcerer's Apprentice"), 11 (*En Suspens*, "Suspended") and 12 (*Entrelacs*, "Interlacing"). However, in only one of these (Etude 10) do the two hands ever occupy the same scale area; in the rest, the hands have separate key signatures, and the sum of the scales for each hand always yield the full chromatic gamut. In *Désordre* this is the white-note set in the RH against the black notes in the LH; in *Galamb Borong* (Fig. 6.1) the whole-tone scale 1 in the RH against 0 in the LH; in *En Suspens* the hands exchange the white-note set and [0, 1, 3, 5, 6, 8, 10] (the D-flat major scale); and *Entrelacs* uses the same scale areas (and similar alternations of hand arrangement) as *En Suspens*, though the scales are missing a PC (the white-note set missing the C and the D-flat major scale the F, meaning the scales have no shared PCs). Of these four etudes, three (Nos. 1, 7 and 12) thus use scalar complements that add up to the full 12 PCs; only one of these scales (the RH scale in *Désordre*) is diatonic, strictly-defined; and in none of them is there any sort of tonal anchoring. *En Suspens*, an etude Ligeti called "almost a jazz piece," comes closest to a more functional treatment of the diatonic scale area, though it is ambiguous throughout with no clearly-defined tonal implication.

Of these etudes, perhaps it is *Der Zauberlehrling* that most clearly foreshadows the procedures of Book 3. While much of the piece uses textures that are chromatically saturated (either through chromatic motion or the superimposition of complementary scale areas in the two hands, as in the etudes mentioned above) a little more than the first third of the etude (the opening 47 bars out of 118) use the white-note set exclusively; bars 55–65 are a steady descent down the whole-tone scale 1; and a climax at bar 96 is approached by passagework that first briefly passes through the acoustic scale at bars 88–89 before seven bars of nothing but the black-note set. However, here too there is no tonality embedded in the scale areas; they are rather a kind of ladder for the *perpetuum mobile* rendering of Ligeti's signature micropolyphony, which, instead of moving in chromatic (or even smaller) steps, now uses larger, diatonic ones (Fig. 6.2).

Thus in these earlier etudes, Ligeti largely manages to avoid implications of tonality that the diatonic scale alone would bring, and this context of Ligeti's earlier use of the scale is an important one given that it highlights the unique nature of the procedures in Book 3. The unprecedented macroharmonic restrictions in *White on White* serve as a bridge to the increasingly tonal processes of these final etudes—his arguably most "postmodern" works. While Ligeti's compositions of the 1980s onwards indeed marked something of a clean break from his previous compositional period, the convenient grouping of all of these works under the umbrella of a late style has meant that the continuing stylistic evolution in his final years has largely been overlooked; a study of Book 3 of the etudes reveals a novel direction at the end of his oeuvre, with a new kind of focus and synthesis of scale and harmony.

This book set out to shed light on this arguably defining feature of Book 3—the treatment of macroharmony—using several kinds of graphical analyses. While the various merits and shortcomings of this approach have been explored in the chapters on each etude, it is worth concluding by very briefly observing how certain explicitly tonal works are presented by these graphs, broadening the context for the interpretation of Book 3 and furthering the discussion on the relative efficacy of this type of analysis in other musical contexts.

For the sake of comparison, the two pieces chosen are both piano etudes: Chopin's Etude in A-flat major Op. 10 No. 10 (Fig. 6.3) and Scriabin's Etude in D-flat major (Fig. 6.4). They are both good candidates for this kind of data processing as these are *perpetuum mobile* pieces: the two hands in both works largely adhere to a constant RU (8th-notes in the Chopin, 16ths in the Scriabin), yielding a neat and reflective distribution of data on the x-axis. Furthermore, despite some accentuation, the dynamic levels vertically across the texture are mostly even. The two works are also useful for comparison between each other. While they are both clearly tonal (and use the framework of a thematic RH vs. broken-chord LH) the Scriabin etude has a more chromatic and experimental harmonic language; the extent

Fig. 6.2 Etude 10 (*Der Zauberlehrling*), bars 88–95, score. © 2005 Schott Music, Mainz—Germany

to which these differences are reflected in the graphs will be particularly revealing. Figures 6.3 and 6.4 present these works in a condensed analysis, comparing the Co5 PC, 7-PC correlation strength and ±6 force comparison graphs.

Figures 6.3 and 6.4 help to illustrate the fact that these graphs are not designed to express or quantify tonality, but rather the use of scale. For example, graphs for *White on White* (or indeed, the majority of any of the etudes of Book 3) look considerably cleaner and more diatonic than those for the Chopin or Scriabin etudes—even though the piece is nowhere near as tonal. While one can on the other hand easily imagine the graphs for a certain atonal work showing chromatic saturation, weak correlation strength and a lack of diatonic force, such results may also show up for a tonal piece, as shown by Fig. 6.4 for the Scriabin etude.

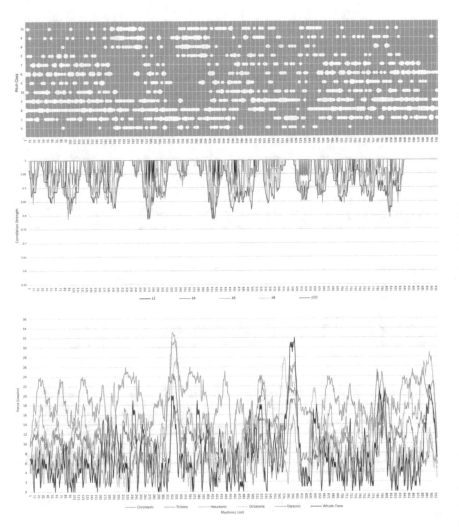

Fig. 6.3 Chopin Etude Op. 10 No. 10, comparison of three graph types (Co5 PC, 7-PC correlation strength and ±6 force comparison graph)

For the Chopin (Fig. 6.3), the Co5 PC graph shows the predominance of the PCs of the A-flat major scale (the full span of PC 1 up through 7). Where this scale is not fully present, other diatonic scales can also be identified by contiguous stretches of PCs, such as the E major (PC 9 up through 3) around RUs 200–260 or the A major (PC 4 up through 8) around RUs 340–400. Accordingly, the seven-PC correlation strength and diatonic force showings are generally strong. However, there are surges of other forces in a few moments where the figuration in the RH departs from the (usually diatonic) stepwise movement (as in the opening, Fig. 6.5) and instead parks on one step or harmony for at least the length of a measure; the nature of the harmony in question will determine which force comes to the fore. For example, the exclusive

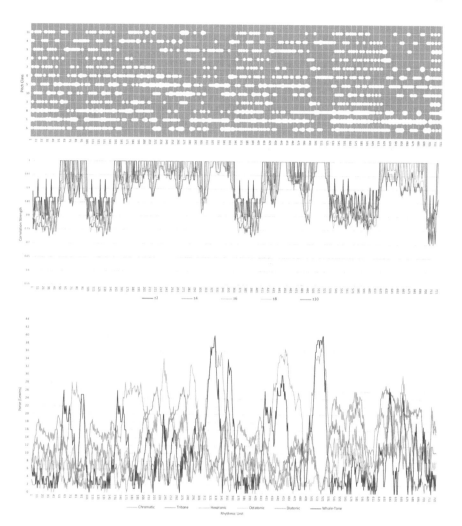

Fig. 6.4 Scriabin Etude Op. 8 No. 10, comparison of three graph types (Co5 PC, 7-PC correlation strength and ±6 force comparison graph)

focus of bars 27 and 28 on the PCs of the D-flat major and D-flat minor triads (Fig. 6.6) creates a perfect hexatonic subset, causing a surge in the force showing from Fourier Balance 3 (peaking at RU 320). It is worth keeping in mind that, given the nature of the balances, a major chord in fact yields more force on balance 3 (2.24 Lw) than on balance 5 for the diatonic force (1.93 Lw).

By comparison, the Co5 PC graph for the Scriabin etude shows no discernible primary scale area or key. The majority of the work is chromatically saturated, and passages that are not show the predominance of PC sets that are not diatonic. The correlation strength graph moves in synchrony, with weak showings in passages of chromatic saturation and a recovery in areas of more restricted dominant PC sets. As

Fig. 6.5 Chopin Etude Op. 10 No. 10, bars 1–8 (RUs 1–97), score. G. Schirmer, New York, 1895

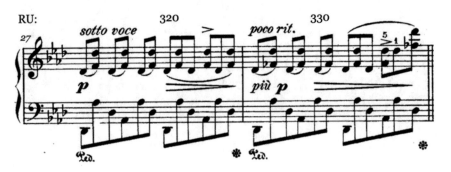

Fig. 6.6 Chopin Etude Op. 10 No. 10, bars 27–28 (RUs 314–337), score. G. Schirmer, New York, 1895

with the Chopin, peaks in the force graph are also coordinated with the correlation strength—but fascinatingly, in rather the opposite manner, with a weak but present diatonic force in areas of chromatic saturation, and a surge in other forces (primarily the hexatonic and whole-tone) when the texture clears. The opening of the work (Fig. 6.7) contains both kinds of moments. Given the chromatic saturation of the RH in the first eight bars, it is the LH-texture that will largely determine the force—and given the clean diatonicity and focus on fifths, the diatonic comes to the fore (albeit somewhat weakly). As soon as the RH begins to outline particular harmonies, other forces surge (much like in the Chopin)—in particular the whole-tone and hexatonic,

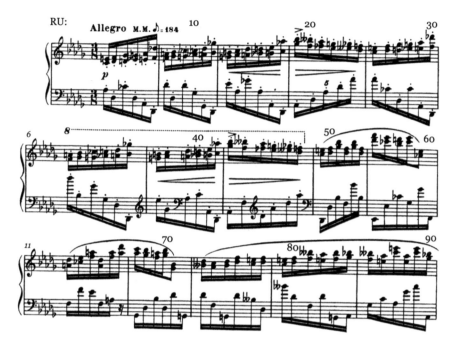

Fig. 6.7 Scriabin Etude Op. 8 No. 10, bars 1–15 (RUs 1–90), score. Muzgiz, Moscow, 1947

Fig. 6.8 Scriabin Etude Op. 8 No. 10, bars 5–8 (RUs 25–48), PCs rearranged in register

given that the etude focuses heavily on augmented triads and parallel major-third motion in either tones (as in bar 11) or semitones (bars 12–13).

In the Co5 PC graph, one can glean a slight predominance of the PCs 1, 5, 6 and 8—the PCs of the D-flat major triad (1, 5 and 8) plus the subdominant scale degree (6), thus hinting at the presence of some kind of tonal anchoring. However, one could change the P-space arrangement of each PC in the Scriabin (namely moving each note to a different octave, while maintaining the rhythm and all other features) in a way that would eliminate (or at least make barely perceptible) the sense of tonality, key or scale area, even for the more diatonic passages as shown in Fig. 6.8 (which, for the sake of illustration, includes a number of enharmonic respellings); while one could also do this for the Chopin, the presence of diatonic scale areas would still come to the fore even if a sense of tonality were weakened.

This is an important illustration that, *ceteris paribus*, registration alone (broadly defined, which importantly includes voice-leading) can determine the extent to which we perceive a texture as tonal or not. Bars 5–8 from Fig. 6.7 are undoubtedly tonal while those in Fig. 6.8 are not; yet the two passages would yield identical results on any graphs that use PC- instead of P-space (which, in this book, are all the graph types apart from the piano roll). As sensitive as the graphs may be to the scalar composition of macroharmony, they are not able to reflect the harmonic quality that results from the arrangements of those PCs in P-space—an important consideration in the adoption of this analytical technique in different musical contexts.

Nevertheless, these graphs are veritable x-rays for textural composition with regards to the use of scale. Not only do they present several layers of information for individual passages, but also respond dynamically to how this idea develops over the course of an entire work; more broadly, they track and to some extent quantify a listener's perception of changing levels of stability and instability as a result of these macroharmonic processes, shedding light on the manner in which large-scale structures and dramatic arcs are built through macroharmonic juxtaposition. In the present analysis, the graphs help show the radical nature of Ligeti's use and treatment of scale area in the etudes of Book 3 (especially when considered relative to his earlier etudes, or even much of the preceding canon in the genre, as the above discussion shows) and the remarkable manner in which macroharmony is used to craft their narratives—a set of conclusions that shall hopefully encourage further investigations into these works.

Appendix A: Scrolling Videos

A central consideration in the discussion of macroharmony throughout this book is the question of how we as listeners perceive the textures analyzed. While the printed graphs manage to reflect this to a certain extent, they are best experienced when synchronized to a recording of the music. In the four videos below (Links A.1, A.2, A.3 and A.4), the three most frequently used graph types—the Co5 PC, seven-PC correlation strength and ±6 force comparison graphs, aligned with a notation row above them—scroll vertically along with the progress of a software-produced recording of each etude.

While the videos for Etudes 16–18 show each etude in full, the one for Etude 15 focuses on the conclusion, as this is the only part of the work that is addressed in depth in the analysis. As with the graphs, the notation rows are devoid of dynamics, extra stems for voice assignments, expression and tempo marks and so forth. However, these performance indications are reproduced in the recordings, and the speed of the vertical scrolling changes depending on the tempo.

When watching the videos it is important to keep in mind that the spans anticipate events happening beyond the current timepoint. For example, a ±6 span at RU 408 includes data from RU 402 to RU 414. This means that the data at any timepoint reflects musical events that have not yet happened in the recording (as well as those that have already happened, though this is more straightforward to follow given our memory of what we have just heard).

Link A.1 Etude 15, RUs 448–620, scrolling video. https://youtu.be/9SEOM2 xRrrA

Link A.2 Etude 16, RUs 1–598, scrolling video. https://youtu.be/G9HrIpuBOsg

Link A.3 Etude 17, RUs 1–943, scrolling video. https://youtu.be/ZwiVqZ9QI_I

Link A.4 Etude 18, RUs 1–196, scrolling video. https://youtu.be/807lwb3rDlI

N. Namoradze, *Ligeti's Macroharmonies*, Computational Music Science,
https://doi.org/10.1007/978-3-030-85694-6

Appendix B: Maximal Forces on Fourier Balances

Table B.1 shows the force readings for the maximal sets (i.e. the PC sets that produce the maximum possible force on each balance) for balances 1 through 6. As mentioned in Chap. 4, these do not form a readily discernible sequence. However, it begins to make a little more sense once one visualizes the balances. For example, the six PCs of the maximal set for balance 6 all move in the same direction and thus compound perfectly, whereas on balance 5 they are spread out in different directions (Fig. B.1). The narrower the geometric span around the circle that the maximal set occupies, the stronger the force.

In algebraic terms, the maximal force m on any balance b can be expressed as follows:

$$m(b) = \sqrt{\left(\sum_{n}^{n+5} \sin\left(\frac{\pi b}{6}\left(2n - \frac{n(b_{mod6})}{b}\right)\right)\right)^2 + \left(\sum_{n}^{n+5} \cos\left(\frac{\pi b}{6}\left(2n - \frac{n(b_{mod6})}{b}\right)\right)\right)^2}$$

(B.1)

Equation (B.1) uses any six integers n to $n+5$ (the choice of n does not affect the final result). The equation has broadly recognizable features compared to the one for calculating the force of any PC set, as discussed in the introductory chapter (such as

Table B.1 Force readings for maximal sets on each balance

Balance	Maximal set	
	Set	Force (Lw)
1	[012345]	3.86
2	[012678]	4.00
3	[014589]	4.24
4	[0134679T]	4.00
5	[024579]	3.86
6	[02468T]	6.00

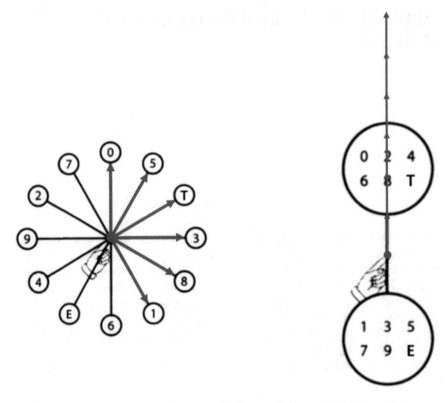

Fig. B.1 Comparison of forces applied by maximal sets on balances 5 and 6 (forces in the same direction are superimposed)

the Pythagorean-theorem formulation and the summing of the x- and y-components of each PC force arrow). This equation however yields the maximum force on any balance b from simply knowing the balance number, without requiring the input of particular PCs.

This is made possible by an algebraic process that can in fact produce the maximal set for any balance b—rather than simply observing each balance and counting the PCs within a 180° span. The maximal set M for balance b can be derived in Eq. (B.2) for any range of input x:

$$M_b = \left\{ \left(\frac{x(b_{mod6})}{b} + \left(x\frac{12}{b} \right)_{mod\,12} \right) \cap \mathbb{Z} \right\} \qquad \text{(B.2)}$$

Any integer result for any input of x will be one of the PCs for M, the maximal set. The equation yields the following sets (ordered by PC and ignoring repetitions):

$M_1 = \{0, 1, 2, 3, 4, 5\}$
$M_2 = \{0, 1, 2, 6, 7, 8\}$
$M_3 = \{0, 1, 4, 5, 8, 9\}$
$M_4 = \{0, 1, 3, 4, 6, 7, 9, 10\}$
$M_5 = \{0, 1, 3, 5, 8, 10\}$
$M_6 = \{0, 2, 4, 6, 8, 10\}$

Rather than fishing for integer results (Table B.2) and reordering them, one can instead use an equation that will more straightforwardly yield the maximal set (a somewhat preferable process for the deriving of the maximum force):

$$M_b = \left\{ \left(2x - \frac{x(b_{mod6})}{b} \right) mod\, 12 \,\middle|\, x \in \mathbb{Z}, 0 \le x < 6 \right\} \qquad (B.3)$$

Equation (B.3) derives the maximal set as relating to the steps from one panhandle to the next—thus requiring that x be restricted to integers. It also requires a more rigid conception of the maximal set as a 6-PC set: as the 12 PCs are evenly distributed (in different configurations) around the 360° of each balance, 6 of them should form the maximal 180° span. These two features will have implications for the results for balances 4 and 5:

$M_1 = \{0, 1, 2, 3, 4, 5\}$
$M_2 = \{0, 1, 2, 6, 7, 8\}$
$M_3 = \{0, 1, 4, 5, 8, 9\}$
$M_4 = \{0, 1, 3.5, 6, 7, 9.5\}$
$M_5 = \{0, 1, 3.2, 5.4, 7.6, 9.8\}$
$M_6 = \{0, 2, 4, 6, 8, 10\}$

The input for x can in fact range through all the integers—however, the cycle repeats every six integers as shown by Table B.3, so only six consecutive integers are needed for our purposes.

As shown in Table B.3, this equation gives the maximal set in scalar order. The apparently peculiar results for balances 4 and 5 are explained as soon as one reimagines these results on the balances themselves.

Figure B.2 shows that the results for M_4 are simply a 6-PC rendering of the octatonic set, where 3.5 is the equivalent of PCs 3 *and* 4: it is halfway between the two, and given the spatial arrangement produces the same force as 3 and 4 together. The same applies for 9.5 as a combination of 9 and 10. If we in fact imagine the other adjacent PC pairs as single "PC steps"—namely 0.5 and 6.5—we would end up with four PC steps applying placed directionally opposite the set 025E, further illustrating how the octatonic set applies equal and opposite force to its four-PC, diminished-seventh chord complement on balance 4.

The issue with balance 5 is a little more complex. As mentioned above, the steps from one panhandle to the next are a key feature in this formula. In balances 1, 2, 3, 4 and 6, a (clockwise) step from one panhandle to the next is always 1, modulo 12/b. For example, on balance 4 one can take any PC on any panhandle, any PC

Table B.2 Results for Eq. (B.2), using the set of natural numbers (including 0) for x

x	b					
	1	2	3	4	5	6
0	0	0	0	0	0	0
1	1	7	5	4	3.4	2
2	2	2	8	6.5	5.6	4
3	3	6	1	9	7.8	6
4	4	1	4	1	10	8
5	5	8	9	3.5	0.2	10
6	0	0	0	6	2.4	0
7	1	7	5	10	5.8	2
8	2	2	8	0.5	8	4
9	3	6	1	3	10.2	6
10	4	1	4	7	0.4	8
11	5	8	9	9.5	2.6	10
12	0	0	0	0	4.8	0
13	1	7	5	4	8.2	2
14	2	2	8	6.5	10.4	4
15	3	6	1	9	0.6	6
16	4	1	4	1	2.8	8
17	5	8	9	3.5	5	10
18	0	0	0	6	7.2	0
19	1	7	5	10	10.6	2
20	2	2	8	0.5	0.8	4
21	3	6	1	3	3	6
22	4	1	4	7	5.2	8
23	5	8	9	9.5	7.4	10
24	0	0	0	0	9.6	0
25	1	7	5	4	1	2
26	2	2	8	6.5	3.2	4
27	3	6	1	9	5.4	6
28	4	1	4	1	7.6	8
29	5	8	9	3.5	9.8	10
30	0	0	0	6	0	0
31	1	7	5	10	3.4	2
32	2	2	8	0.5	5.6	4
33	3	6	1	3	7.8	6
34	4	1	4	7	10	8
35	5	8	9	9.5	0.2	10

on the next panhandle, and their sum mod 3 will always be 1. Thus the movement from PC p to PC $p+1$ is always a matter of moving to the (immediately) next panhandle: PC steps are equal to panhandle steps. Therefore, when counted from

Table B.3 Results for Eq. (B.3), using the set of natural numbers (including 0) for x

x	b					
	1	2	3	4	5	6
0	0	0	0	0	0	0
1	1	1	1	1	1	2
2	2	2	4	3.5	3.2	4
3	3	6	5	6	5.4	6
4	4	7	8	7	7.6	8
5	5	8	9	9.5	9.8	10
6	0	0	0	0	0	0
7	1	1	1	1	1	2
8	2	2	4	3.5	3.2	4
9	3	6	5	6	5.4	6
10	4	7	8	7	7.6	8
11	5	8	9	9.5	9.8	10
12	0	0	0	0	0	0
13	1	1	1	1	1	2
14	2	2	4	3.5	3.2	4

Fig. B.2 Fourier Balance 4 with forces from PC steps 0, 1, 3.5, 6, 7 and 9.5 (forces in the same direction are superimposed)

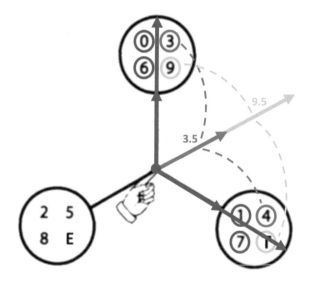

0 the panhandle steps will simply equal the PC. For each of these balances, 12/b yields the number of panhandles (balance 6 has two panhandles, balance 3 has four etc.) and thus the number of panhandle steps needed to make a full revolution.

Balance 5 is an exception: it is the only balance where PC steps do not equal panhandle steps. Instead, in order to move from PC p to PC $p+1$, one has to in fact move five panhandles. Thus the movement from one panhandle to the next is 0.2 PC steps. (It is the only balance where other PCs are mapped *between* the space from p to $p+1$.) As soon as for example 3.2 is interpreted as a PC step rather than a PC

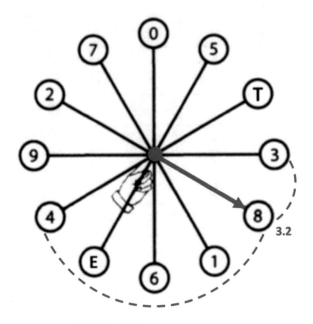

Fig. B.3 Fourier Balance 5 with PC step 3.2

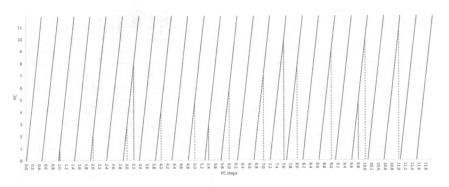

Fig. B.4 PC steps against PCs on Fourier Balance 5 (each PC step mapped with black dotted lines, and the PC steps 0, 1, 3.2, 5.4, 7.6, 9.8 with orange ones)

(which in other balances would be identical), namely 1/5 of the distance between PCs 3 and 4, the PC it refers to becomes immediately apparent (Fig. B.3).

Thus, PC step 3.2 is in fact PC 8, 5.4 is 3, 7.6 is 10 and 9.8 is 5—yielding the same {0, 1, 3, 5, 8, 10} set as in the first formula. This is also demonstrated by the graph in Fig. B.4, which maps the PC and PC steps.[1]

---

[1] One can apply this procedure to the results of Eq. (B.2) (Table B.2) as well, revealing a repeating {0, 1, 8, 3, 10, 5} pattern.

In order to convert the PC steps to PCs on balance 5, one can first convert them to panhandle steps by multiplying them by 5, and then again by 5 to reflect the number of PCs traveled on each panhandle step, modulo 12. Thus, PC steps s become PCs by the process $(5^2 s)_{\text{mod}12}$.

However, whether one uses PC steps (namely 0, 1, 3.2, 5.4, 7.6, 9.8) or PCs (0, 1, 3, 5, 8, 10) has no effect on Eq. (B.1) (repeated below) for calculating the maximum possible force on the balance—they yield the same result.

$$m(b) = \sqrt{\left(\sum_{n}^{n+5} \sin\left(\frac{\pi b}{6}\left(2n - \frac{n(b_{mod6})}{b} \right) \right) \right)^2 + \left(\sum_{n}^{n+5} \cos\left(\frac{\pi b}{6}\left(2n - \frac{n(b_{mod6})}{b} \right) \right) \right)^2}$$

This equation can also be molded to apply to other settings. For example, it can give the maximum force on any balance for other PC universes—not just 12-PC ones. In fact, given that the maximum force is derived from the forces applied by all PCs within a 180° span on any balance, one can apply yet another level of abstraction—finding the force exerted by *any* span of PCs (where each and every PC in that portion appears once). Namely, if 360° is represented by 1, the portion y is expressed as a fraction or decimal—such as a quarter of the total span, which would be $\frac{1}{4}$ or 0.25. Thus, the following equation yields a way to derive the force of any span on any balance in any c-PC universe:

$$m(b) =$$

$$\sqrt{\left(\sum_{n}^{n+cy-1} \sin\left(\frac{\pi b}{cy}\left(2n - \frac{n(b_{mod cy})}{b} \right) \right) \right)^2 + \left(\sum_{n}^{n+cy-1} \cos\left(\frac{\pi b}{cy}\left(2n - \frac{n(b_{mod cy})}{b} \right) \right) \right)^2}$$

$$(\text{B.4})$$

Bibliography

Scores

Chopin, F. 1895. *Etudes for Piano, Op. 10*. New York: G. Schirmer.
Ligeti, G. 1982. *Trio for Violin, Horn and Piano*. Mainz: Schott.
— 1985. *Études pour piano, premier livre*. Mainz: Schott.
— 1988. *Piano Concerto*. Mainz: Schott.
— 1992. *Violin Concerto*. Mainz: Schott.
— 1994. *Études pour piano, deuxième livre*. Mainz: Schott.
— 2000. *Síppal, dobbal, nádihegedűvel*. Mainz: Schott.
— 2001a. *Études pour piano, troisième livre*. Mainz: Schott.
— 2001b. *Hamburgisches Konzert*. Mainz: Schott.
Pärt, A. 1976. *Für Alina*. Vienna: Universal Edition.
Scriabin, A. 1947. *Etudes for Piano, Op. 8*. Moscow: Muzgiz.

Sources

Amiot, E. 2016. *Music Through Fourier Space: Discrete Fourier Transform in Music Theory*. Cham: Springer.
Barone, A. 1995. Richard Wagner's "Parsifal" and the Theory of Late Style. *Cambridge Opera Journal*. Vol. 7 No. 1: 37–54.
Bauer, A. 2010. Philosophy Recomposed: Stanley Cavell and the Critique of New Music. *Journal of Music Theory*. Vol. 54 No. 1: 75–90.
— 2011. *Ligeti's Laments: Nostalgia, Exoticism and the Absolute*. Farnham: Ashgate Publishing.
Bernard, J. W. 1999. Ligeti's Restoration of Interval and its Significance for His Later Works. *Music Theory Spectrum*. Vol. 21 No. 1: 1–31.
Bossin, J 1984. György Ligeti's New Lyricism and the Aesthetic of Currentness: The Berlin Festival's Retrospective of the Composer's Career. *Current Musicology*. 37/38: 233–239.
Boros, J. 1994. Why Complexity? *Perspectives of New Music*. Vol. 32 No. 1: 90–101.
— 1996. A Response to Lerdahl. *Perspectives of New Music*. Vol. 34, No. 1: 252–258.
Browne, R. 1981. Tonal Implications of the Diatonic Set. *In Theory Only*. 5: 3–12.
Callender, C. 2017. Complementary Collections in Ligeti's *Désordre*. Accessed July 21, 2020. https://cliftoncallender.com/img/pdfs/CallenderMCM2017.pdf.

© The Author(s), under exclusive license to Springer Nature Switzerland AG 2022
N. Namoradze, *Ligeti's Macroharmonies*, Computational Music Science,
https://doi.org/10.1007/978-3-030-85694-6

Carey, N. 2002. On Coherence and Sameness, and the Evaluation of Scale Candidacy Claims. *Journal of Music Theory*. Vol. 46, No. 1/2 (Spring–Autumn): 1–56.

Clough, J. 1979. Aspects of Diatonic Sets. *Journal of Music Theory*. 23: 45–61.

Clough, J, and G. Myerson. 1985. Variety and Multiplicity in Diatonic Systems. *Journal of Music Theory*. 29: 249–270.

Cuciurean, J. D. 2000. *A Theory of Pitch, Rhythm, and Intertextual Allusion for the Late Music of György Ligeti*. PhD diss., State University of New York at Buffalo.

Drott, E. 2003. The Role of Triadic Harmony in Ligeti's Recent Music. *Music Analysis*. Vol. 22 No. 3, 283–314.

Floros, C. 1996. *Jenseits von Avantgarde und Postmoderne*. Vienna: Verlag Lafite.

Egmond, R. and D. Butler, 1997. Diatonic Connotations of Pitch Class Sets. *Music Perception: An Interdisciplinary Journal*. Vol. 15 No. 1: 1–29.

Gauldin, R. 1983. The Cycle-7 Complex: Relations of Diatonic Set Theory to the Evolution of Ancient Tonal Systems. *Music Theory Spectrum*. 5: 39–55.

Griffiths, P. 2006. Music in the Modern-Postmodern Labyrinth. *New England Review*. Vol. 27 No. 2, 96–113.

Hagen, D. 1992. Everything old is new again. *Contemporary Music Review*. Vol. 6 No. 2: 51–52.

Harbison, J. 1992 Symmetries and the "New Tonality." *Contemporary Music Review*. Vol. 6 No. 2: 71–79.

Hentschel, F. 2006. Wie neu war die "Neue Einfachheit"? *Acta Musicologica*. Vol. 78 No. 1: 111–131.

Hillier, P. 1997. *Arvo Pärt*. Oxford: Oxford University Press.

Hinson, M. 2000. *Guide to the Pianist's Repertoire*. 3rd edn. Bloomington: Indiana University Press.

Howard, R. 2014. *Musical Rhetoric, Narrative, Drama and Their Negation in Morton Feldman's Piano and String Quartet*. PhD Diss., The Graduate Center, City University of New York. CUNY Academic Works.

Ince, K. 1992. Emancipation of tonal sonorities. *Contemporary Music Review*. Vol. 6 No. 2: 49–50.

Knittel, K. M. 1998 Wagner, Deafness, and the Reception of Beethoven's Late Style. *Journal of the American Musicological Society*. Vol. 51, No. 1: 49–82.

Kompridis, N. 1993. Learning from Architecture: Music in the Aftermath to Postmodernism. *Perspectives of New Music*. Vol. 31 No. 2: 6–23.

Kresky, J. 1986. What's on the Horizon? *Perspectives of New Music*. Vol. 24 No. 2: 434–38.

Lennon, J. 1992. The Daedalian Factor: Tonality, Atonality, or Musicality? *Contemporary Music Review*. Vol. 6 No. 2: 23–25.

Lerdahl, F. 1996. Tonality and Paranoia: A Reply to Boros. *Perspectives of New Music*. Vol. 34 No. 1: 242–251.

Lewin, D. 2001. Special Cases of the Interval Function between Pitch-Class Sets X and Y. *Journal of Music Theory*. Vol. 45 No. 1: 1–29.

Ligeti, G. 1978. "On Music and Politics." *Perspectives of New Music*. Vol. 16 No. 2, 19–24.

— 1990. Ma position comme compositeur aujourd'hui. *Contrechamps*. Vol. 12/13: 8–9.

— 1996. Liner note to *György Ligeti Edition Vol. 3: Works for Piano*. Compact Disc. Sony. SK 62308.

Ligeti, G., P. Várnai, J. Häusler and C. Samuel. 1983. *Ligeti in conversation*. London: Eulenburg Books.

Livingston, C. 2010. A Leap of Faith: Composing in the Wasteland of Postmodernism. *Tempo*. Vol. 64 No. 253: 30–40.

Marczi, M. 2008. *Ligeti György Zongoraetüdjei*. Liszt University of Music Budapest, Hungary Doctoral Dissertation.

Pellegrino, C. 2002. Aspects of Closure in the Music of John Adams. *Perspectives of New Music*. Vol. 40 No. 1: 147–75.

Perle, G. 1992. Symmetry, the twelve-tone scale, and tonality. *Contemporary Music Review*. Vol. 6 No. 2: 81–96.

Quinn, I. 2007. General Equal-Tempered Harmony: Parts 2 and 3. *Perspectives of New Music*. Vol. 25 No. 1 (Winter): 4–63.

Quinnett, L. 2014. Harmony and Counterpoint in the Ligeti Etudes, Book I: An Analysis and Performance Guide. DMA Diss., Florida State University. FSU Digital Library.

Rigoni, M. 1998. Musique et postmodernité: pour un état des lieux. *Musurgia*. Vol. 5 No. 3. 109–122.

Sabbe, H. 1979. György Ligeti – illusions et allusions. *Interface*. Vol. 8 No. 11, 11-34.

Santa, M. 2000. Analyzing Post-Tonal Music: A Modulo 7 Perspective. *Music Analysis*. Vol. 19 No. 2 (July): 167–201.

Scherzinger, M. 2006. György Ligeti and the Aka Pygmies Project. *Contemporary Music Review*. 25/iii: 227–62.

— 2008. Musical Modernism in the Thought of "Mille Plateaux," and Its Twofold Politics. *Perspectives of New Music*. Vol. 46 No. 2: 130–58.

Schmid, W. 1992. World Music in the Instrumental Program. *Music Educators Journal*. Vol. 78 No. 19: 41–45.

Schoenberg, A. 1926. "Opinion or Insight." In *Style and Ideas: Selected Writings of Arnold Schoenberg*, ed. L. Stein, tr. L. Black, 258–263. Berkeley: University of California Press (1984).

Schwarz, K. 1996. *Minimalists*. London: Phaidon Press.

Searby, M. 1997. Ligeti the Postmodernist? *Tempo, New Series*. 199: 9–14.

— 2001. Ligeti's 'third way': 'Non-atonal' elements in the Horn Trio. *Tempo, New Series*. 216: 17–22.

Shaffer, K. 2011. "Neither Tonal Nor Atonal"?: Harmony and Harmonic Syntax in György Ligeti's Late Triadic Works. PhD Diss., Yale University.

Spitzer, M. 2006. *Music as Philosophy: Adorno and Beethoven's Late Style*. Bloomington: University of Indiana.

Steinitz, R. 1996. The Dynamics of Disorder. *The Musical Times*. Vol. 137, No. 1839: 7–14.

— 2003. *György Ligeti: Music of the Imagination*. London: Faber and Faber.

Straus, J. 2008. Disability and "Late Style" in Music. *The Journal of Musicology*. Vol. 25 No. 1: 3–45.

Szigeti, I. 1983. "A Budapest Interview with György Ligeti." Prepared by Josh Ronsen. Accessed July 25, 2020. http://ronsen.org/monkminkpinkpunk/9/gl4.html.

Szitha, T. 1995 A Conversation with György Ligeti. *Hungarian Music Quarterly*. Vol. 3 No. 1: 14–17.

Talgam, I. 2019. *Performing Rhythmic Dissonance in Ligeti's Piano Études, Book 1: A Perception-Driven Approach and Re-notation*. DMA diss., The Graduate Center, City University of New York. CUNY Academic Works.

Taylor, S. 1994. *The Lamento Motif: Metamorphosis in Ligeti's Late Style*. DMA Dissertation, Cornell University.

— 2003. Ligeti, Africa and Polyrhythm. *The World of Music*. Vol. 45 No. 2: 83–94.

— 2012. Hemiola, Maximal Evenness, and Metric Ambiguity in Late Ligeti. *Contemporary Music Review* 31. 2/3: 203–20.

Thomas, G. 1993. New Times: New Clocks. *The Musical Times*. Vol. 134 No. 1805: 376–379.

Toop, R. 1999. *György Ligeti*. London: Phaidon Press.

Tymoczko, D. 2011. *Geometry of Music*. Oxford: Oxford University Press.

Wierzbicki, J. 2007. Reflections on Rochberg and "Postmodernism." *Perspectives of New Music*. Vol. 45, No. 2: 108–132.

Printed in the United States
by Baker & Taylor Publisher Services